The Secret Life of Chemicals

Alfred Poulos

The Secret Life of Chemicals

 Springer

Alfred Poulos
Thornbury, VIC, Australia

ISBN 978-3-030-80340-7   ISBN 978-3-030-80338-4   (eBook)
https://doi.org/10.1007/978-3-030-80338-4

This Springer imprint is published by the registered company Springer Nature Switzerland AG
The registered company address is: Gewerbestrasse 11, 6330 Cham, Switzerland

Acknowledgements

I would like to acknowledge the help, support and encouragement of my wife Jan and my daughters Danielle and Christianne. My hope is that my two grandsons, Oliver and Lewis, and their future children and grandchildren will live in a less polluted world.

Contents

1 Introduction 1
 References 3

2 Pesticides in Our Food 5
 What Are the Amounts of Pesticides Used Around the World? 5
 Why Use Pesticides in Agriculture? 5
 Some Problems with Pesticides 6
  How Do Pesticides Work? 7
 How Poisonous Are Pesticides to Humans? 9
 How Is Toxicity Resulting from Chronic Exposure Detected? 10
 How Do We Measure the Toxicity of Pesticides? 10
 Assessment of Toxicity of Pesticides in Humans 11
 Regulation of Pesticide Use 12
 Monitoring of Pesticide Levels in Food 13
 What Sorts of Levels of Pesticides Are Present in Our Food? 13
 Is Government Regulation of Pesticide Use Adequate to Protect
 Us from Harm? 15
 Summary – Pesticides (General) 17
 What Is the Evidence for Toxicity of Pesticides in Humans? 17
  Acute Toxicity 17
  Chronic Effects 18
 Diseases Caused by Occupational Exposure to Pesticides 19
  Brain Diseases 20
  Birth Defects 20
  Cancer 21
  Known Unknowns? 22
 Diseases Caused by Exposure to Pesticides Outside the Work
 Environment 22
 Varying Susceptibility 24
  Children 24

The Unborn Baby   26
Adults Who Increased Susceptibility   26
Toxicity of Other Components in Pesticide Formulations   28
What Does All This Mean?   28
Summary – Pesticides and Our Health   29
References   30

3   The Plastics Revolution   33
How Are Plastics Made?   34
What Types of Plastic Are There?   34
What Are Degradable Plastics?   36
What Does Plastic Contain?   37
Why Are There Additives in Plastic?   37
Nanomaterials   38
What Happens When Plastic Degrades?   38
How Can the Chemicals in Plastics Be Taken Up into
Our Bodies?   39
Are Plastic Components Harmful?   41
What Does All This Mean?   44
References   44

4   Toxic Metals   49
Coal Combustion as a Major Source of Environmental Heavy Metal
Contamination   49
Lead   50
Sources of Exposure   50
What Are "Safe" Blood Levels of Lead?   52
What Is the Evidence That Lead Is Harmful?   52
Mercury   53
What Are the Industrial Uses of Mercury?   53
Mercury in Food   54
What Are the Concerns About Mercury?   54
Arsenic   55
What Are the Industrial Uses of Arsenic?   55
Arsenic in Ground Water and Coal   56
Arsenic in Food   56
What Are the Concerns About Arsenic?   57
What Does All of This Mean?   58
References   58

5   The Indestructibles   63
Regulation of Industrial Chemicals   63
Polychlorinated Biphenyls (PCBs)   64
Do PCBs and Dioxins Enter the Food Chain?   65
Are the PCBs and Dioxins Taken up into the Body?   65
What Do PCBs and Dioxins Do to Animals?   65

Can PCBs and Dioxins Affect Our Health?..................... 66
PCBs, Dioxins and Health – The Bottom Line.................. 68
    Polybrominated Diphenyl Ethers (PBDEs)................... 68
Do PBDEs Enter the Food Chain?........................... 68
Do PBDEs Get into Our Bodies?............................ 69
What Do PBDEs Do to Animals?............................ 69
Can PBDEs Affect Our Health?............................. 70
What Does All This Mean?................................. 70
References.. 71

6   Air Pollutants.. 75
What Are the Main Pollutants in the Air We Breathe?............. 75
Household Air Pollution................................... 76
What Sorts of Chemicals Occur in PMs?...................... 77
Are the PMs and Volatile Organic Components Taken up into Our
Bodies?.. 77
What Impact Do PMs and Other Pollutants in the Air Have on Our
Health?.. 78
    Gas Pollutants....................................... 78
    PMs.. 79
What Does All This Mean?................................. 81
References.. 82

7   Water Everywhere – But Is it Safe to Drink?................. 85
Reservoir Water... 86
Stormwater.. 87
Graywater... 88
Sewerage.. 88
Tank Water... 89
Bottled Water... 89
Rainwater.. 90
Desalinated Water....................................... 90
Tapwater.. 91
Groundwater... 91
Recycled Water... 92
Biosolids.. 92
How Is Reservoir Water Made Safe for Drinking?............... 92
The Disinfection Process.................................. 95
Chlorination... 95
Other Disinfection Processes.............................. 96
Transport and Distribution of Drinking Water.................. 97
Government Regulation of Drinking Water Quality.............. 97
Monitoring of Water Quality............................... 98
Drinking Water and Disease............................... 99
Naturally Occurring Chemical Contaminants and Disease.......... 99

Industrial, Agricultural, and Other Chemicals in Water and Disease . . .   100
Radioactivity in Water and Disease   101
Plastics in Water and Disease   101
Plumbing Contaminants and Disease   102
Disinfection Byproducts and Disease   102
How Do We Determine Toxicity of Disinfection Byproducts?   103
  How Are We Exposed to the Disinfection Byproducts and Other
  Chemicals in Water?   104
Are Disinfection Byproducts and Other Contaminants Taken
up into Our Bodies?   105
Is There any Evidence That Chlorination Byproducts Cause
Disease in Humans?   106
Occupational Exposure to Disinfection Byproducts and Disease   106
Disinfection Byproducts and Disease   107
What Does This Mean?   108
References   108

8   Paper Manufacture and Use   115
Chemical Byproducts of Paper Manufacture   116
Bleaching of Pulp   117
Non Chemical Processes   118
What Else Is Added to Paper?   118
What Are the Potential Environmental Contaminants Released
During the Manufacture and Disposal of Paper?   118
What Impact Does Paper Manufacture and Disposal Have
on Our Health and Wellbeing?   119
Incineration of Paper and Pollutants   121
Landills and Pollutants   122
Recycling and Pollutants   122
What Does All This Mean?   123
References   124

9   Chemical Exposure in the Workplace   127
Workplaces and Chemical Exposure   128
Occupation and Lung Diseases   129
Mining   129
Occupation and Skin Diseases   130
Agriculture and Horticultural Workplaces   130
Woodworking   131
Food Handling   132
Industrial Activities   132
Exposure to Metals and Disease   132
  Cadmium   132
  Arsenic   133
  Beryllium   134

Indium... 134
Chromium.. 134
Non-metallic Exposure.................................... 134
Formaldehyde... 134
Benzene.. 135
Diacetyl.. 135
Diesel Exhaust.. 136
Other Chemicals...................................... 137
What Does All This Mean?................................. 137
References.. 138

10 Fluorocarbons... 143
Release of Fluorocarbons into the Environment................ 144
Are Fluorocarbons Found in Food and Water?................. 145
Are Fluorocarbons Present in Human Tissues and Body Fluids?...... 145
How Do Fluorocarbons Get into Our Bodies?.................. 146
What Effect Do Fluorocarbons Have on Our Health
and Wellbeing?.. 146
What Are the Effects of Sub-lethal Doses of Fluorocarbons
on Animals?.. 147
What Are the Effects of Sub-lethal Doses on Us?.............. 148
Reproduction and Lactation................................ 149
Semen Quality.. 149
Birth Weight.. 150
Breast Milk... 150
Heart Disease... 150
Kidney Disease.. 150
Thyroid Disease....................................... 151
Gestational Diabetes................................... 151
What Does All This Mean?................................. 152
References.. 153

11 Noise, Chemicals, and Hearing............................ 157
What Is Sound?.. 157
How Is Sound Measured?............................... 158
Sources of Noise...................................... 158
Noise Pollution....................................... 158
The Health Effects of Noise................................ 159
Hearing Loss... 159
Cardiovascular Disease................................. 159
Sleep.. 160
Weight Gain.. 160
Acoustic Neuroma..................................... 160
Tinnitus.. 160

Hearing Loss and Chemicals   161
     What Does This Mean?   161
References   162

12 Radiation Pollutants   165
What Are the Natural Sources of Ionizing Radiation?   166
What Is Radioactivity?   166
What Are the Principal Radioactive Elements Used by Humankind? . . .   167
What Are the Sources of Radioactivity Released into the Environment
by Humankind?   168
Does Our Food and Water Contain Radioactive Substances?   168
What Are the Sources of Non Ionizing Radiation?   169
What Are the Effects of Radioactivity?   170
A Few Words About Radioactive Dose   170
Radioactivity Dosage and Health – Large Amounts   170
Radioactivity Dosage and Health – Small Amounts   171
What Do We Know of the Effects of Small Amounts
of Radioactivity Exposure on Health?   172
Atomic Bomb Survivors   173
Accidents   173
Does Radioactivity Exposure Increase the Risk of Other Diseases?   175
What Do We Know About the Effects of Non-ionising Radiation
on Our Health?   175
Microwave Syndrome   175
Magnetic Resonance Imaging (MRI)   176
Mobile Phones   176
WiFi   176
What Does All This Mean?   177
References   177

13 E-Waste   181
Newer Forms of E-Waste   181
Environmental Release of E-Waste Chemicals   182
What Are the Effects of E-Waste Components on Our Health?   183
What Does This Mean?   185
References   186

14 Do Environmental Chemicals Cause Disease and, If So, How?   189
What Is the Evidence that Chemicals Cause Disease?   191
How Do Environmental Chemicals Enter Our Bodies
and What Happens to Them?   192
The Role of the Liver and Kidney   192
Do Environmental Chemicals Cause Disease?   193
Are There Parts of Our Bodies that Are More Vulnerable to the
Effects of Environmental Chemicals?   194
What Does All This Mean?   195
References   195

15 Genetic Variability and the Risk of Disease- or the Advantages and Disadvantages of Being Different 197
Genetic Polymorphisms and Disease 199
Polymorphisms and Drugs 200
Polymorphisms and Risk of Cancer 201
Polymorphisms and Heart Disease 202
What Does All This Mean? 203
References 204

16 Environmental Chemicals and Our Genes 205
How Are Our Genes Damaged? 206
Do Environmental Chemicals Damage DNA? 206
Is There Evidence That Environmental Chemicals Affect
DNA Structure in Humans? 207
Epigenetics – Non Inherited Changes in DNA 208
What Does All This Mean? 209
References 210

17 Environmental Chemicals and Mitochondria 211
What Does All of This Mean? 213
References 213

18 Environmental Chemicals and the Immune System 215
How Do Environmental Chemicals Affect the Cells of the Immune
System? 217
Is There Any Evidence That Environmental Chemicals Affect the
Immune System of Animals? 219
Is There Any Evidence That Environmental Chemicals Affect the
Immune System of Humans? 219
What Does All This Mean? 220
References 221

**19 Just Because the Amounts Are Small, Does It Mean
They Are Safe?** 223
The New Toxicology Paradigms – Hormesis 225
Synergism 226
Priming 226
What Does All of This Mean? 227
References 228

20 What Can We Do? 229

Index 233

About the Author

Alfred Poulos Born in Australia of parents from the Greek island of Kastellorizo in the Aegean Sea, Prof Alfred Poulos has a PhD from the London University and a law degree from the University of Adelaide. He has worked in universities, research institutes, and hospitals in the UK, the USA, Canada, Sweden, and Australia. He held the position of Chief Medical Scientist at the Women's and Children's Hospital in Adelaide, South Australia, for many years, and is an adjunct professor in the Medical School of the University of Adelaide, a title awarded for his research into genetic diseases, fats, and fat metabolism. He has published over 150 papers in international scientific and medical journals.

He has self-published eight books, including the *Silent Threat, Weight Loss: Navigating the Maze of Strategies, To Be or Not to Be – a Vegetarian, The Benefits of Seafood, Olive Oil –Everything You Want to Know, Organic Food – A Guide for Consumers, Fish Oils – Everything You Want to Know*, and *Milk and Dairy – Friend or Foe*, the latter four books co-authored with Dr Stephen Hardy.

Chapter 1
Introduction

There are many benefits to living in an industrial society but the benefits we receive do come at a price because the waste products of industry are released into the atmosphere, contributing to pollution of the air we breathe, the water we drink, the soils in which we grow our crops, and the oceans which for millennia have provided us with a variety of food. Global warming, smog, and acid rain are well known examples of the effects of the many and varied waste products, mainly chemical pollutants, on our environment.

Global warming is but one example of the potential, and devastating, effects of industrial waste products on the environment. Even governments, perhaps sensing the gradual change in community attitudes, have become increasingly concerned with the possibility of catastrophic consequences of the burning of fossil fuels with the gradual build up of carbon dioxide in the atmosphere followed by climate change. What was once the domain of left leaning, anti-globalisation, and anti-multinational environmental activists, is now mainstream, with talk of Kyoto agreements, cuts in emissions, and carbon credits. Suddenly it is "in" to be green! There is still considerable debate and there are still the skeptics, even some governments, who remain unconvinced that higher atmospheric carbon dioxide levels really do mean increased global warming. However, there does appear to be general agreement, even from the skeptics, that the rise in carbon dioxide levels is a reflection of increased industrial activity.

There is no doubt that the burning of fossil fuels does generate carbon dioxide. If you mix the hydrocarbons, which are simply substances comprising chains of carbon atoms with associated hydrogen atoms, with air containing oxygen, and you provide a little heat, the carbons from the hydrocarbons react with oxygen to form carbon dioxide. From a chemical point of view, carbon dioxide is formed when two oxygen atoms combine with a single carbon. A similar process occurs in our own bodies when the food we eat, also containing carbon atoms, is converted into carbon dioxide which we breathe out. To carry the analogy a little further, the fats we eat are good examples of hydrocarbons in food, and they can generate carbon dioxide.

© The Author(s), under exclusive license to Springer Nature Switzerland AG 2021
A. Poulos, *The Secret Life of Chemicals*,
https://doi.org/10.1007/978-3-030-80338-4_1

If it were only that simple because, under certain conditions, carbon and oxygen can unite in another way to form something else. For example, if there is not enough oxygen to react with all of the carbons, the something else formed is carbon monoxide – a deadly poison. Carbon monoxide can unite with hemoglobin, the red pigment found in the blood, to form carboxyhemoglobin, which then interferes with the capacity of blood to carry vital oxygen to the tissues. If exposure is sufficiently prolonged, death will result. There is little doubt that some carbon monoxide is formed by the burning of fossil fuels but not enough to kill unless in an enclosed space. And, of course, carbon monoxide is not the only other product formed because fossil fuels contain all sorts of other chemicals that can pollute the environment.

While much of the focus has been on the accumulation of carbon dioxide in the atmosphere, little attention has been given to the other consequences of industrial activity of which the burning of fossil fuel is just a single example. Everything we do creates waste- whether it is the manufacture, use and disposal of paper, plastics, electronics components, fertilizers and other agricultural chemicals, or paints to name just a few activities. And, however we get rid of the waste, whether it is through landfills, or through burning, or release into the atmosphere, or into the rivers and oceans, the fact remains that it does not just disappear into some imaginary ether, but goes somewhere and, unless broken down fairly quickly, leaves behind an environmental footprint. Eventually, some of the components of the waste enter the food chain of terrestrial, sea and river animals, and even humans.

The fact of the matter is that planet Earth is finite, although to us it seems so vast and infinite. As industrialisation proceeds, and gathers momentum particularly now that countries with huge populations like India and China develop, the amounts and diversity of waste products must increase. And, the same question that is being asked about one waste product, carbon dioxide, will have to be asked about other waste products. We already know that one of the waste products in oil, sulphur, is released into the air combining with water in the atmosphere and, sooner or later, falling to the ground as acid rain. We also know that the chlorofluoro hydrocarbons, popular refrigerant gases, when released into the atmosphere can interact with one of the components of the atmospheric ozone layer, ozone, affecting its ability to act as a protective shield against the radiation of the sun.

Perhaps not so well appreciated is the fact that there may be other, equally severe but unforeseen consequences, to this gradual accumulation of waste and its various chemical products in the environment. One of these relates to the possible effects of the various waste products, either individually or in combination with others, on our health and wellbeing through their uptake into our bodies and the interference with such basic biological functions as birth, growth, sexual development, and mainte- nance of health [1]. This is no scaremongering because we already know that some chemical waste products released into the environment can enter our bodies via the food that we eat, the water we drink or even the air we breathe. Indeed, analysis of our urine has confirmed that we are exposed to, take up.and then excrete, a complex mixture of chemical pollutants [2]. There are indications that some are harmful to animals affecting, for example, reproduction, and also causing birth defects. But what about us, is there any evidence that they are harmful to us? What we do know is that there are many diseases for which there is no known cause. Examples include literally dozens of autoimmue diseases affecting different parts of our bodies (lupus, thyroiditis, psoriasis, rheumatoid arthritis, type 1 diabetes etc), then there are other diseases such as multiple sclerosis, motor neurone disease, Parkinson's disease, and Alzheimer's disease, autism, macular degeneration and even the adult form of diabetes. Do they contribute to the development of some of these diseases? Is there any evidence for this? And if they do, is there anything we can do about it?

This book is about some of the chemicals we are exposed to as part of our daily life, where they come from and why they are used, their effects on animals and their known, and potential effects on humans. While it is difficult to completely eliminate our exposure to many of these environmental contaminants, it is hoped that, with the information contained in this book, the reader will be in a better position to develop strategies to do this.

References

1. Pruss-Ustun A et al (2011) Knowns and unknowns on burden of disease due to chemicals. Environ Health 10:9. http://www.ncbi.nlm.nih.gov/pmc/articles/PMC3037292/
2. Sexton K et al (2011) Biomarker measurements of concurrent exposure to multiple chemicals and chemical classes. J Toxicol Environ Health A 74:927–942

Chapter 2
Pesticides in Our Food

While most people are aware that growing food involves the use of pesticides, there is little doubt that they would be surprised at the range and diversity of the substances currently used by farmers around the world. Perhaps the best way to get an appreciation of what is being used is to get into a website of one of the agrichemical companies. Alphabetical lists of fruits and vegetables, together with recommended pesticides for each are provided. For example, in one of these sites there are four separate herbicides, five insecticides, and one fungicide recommended for apples, and 26 herbicides suggested for use for barley. A visit to one of the many shops selling agricultural goods and services that are found in most of the small country towns in the developed world is also worthwhile.

What Are the Amounts of Pesticides Used Around the World?

According to the US Environmental Protection Agency (EPA) in a report published in 2011, more than five billion pounds (approximately 2.3 billion kgs or 2.2 million tonnes (UK) of a variety of pesticides were used for the years 2006 and 2007 around the world [1]. Figures for more recent years are not yet available but it is likely that the world wide usage is still in the billions of pounds or kgs. Almost half of the pesticide usage included chlorination chemicals used for the treatment of water. This process, and the chemicals used, is discussed later in an earlier book (The Silent Threat, see Chap. 7).

A. Poulos, *The Secret Life of Chemicals*,
https://doi.org/10.1007/978-3-030-80338-4_2

Why Use Pesticides in Agriculture?

There is nothing new in the use of chemicals to control insects in the growing of food. It dates as far back as ancient Greece and Rome. Homer mentions the use of burning sulphur (a very good fumigant) while Pliny the Elder advises the use of arsenic in controlling insects. The Chinese are believed to have used arsenic-containing substances as insecticides by the sixteenth century, and tobacco extracts, containing nicotine, were used later. Copper arsenite, otherwise known as Paris Green, was introduced into the US to treat the Colorado beetle, and Bordeaux mixture, a copper-containing fungicide, was used to control downy mildew on vines. In the late nineteenth century, iron sulphate and a number of other relatively simple inorganic substances (non-carbon containing) were found to be useful herbicides.

There was a rapid growth in the number and complexity of the pesticides used in the period between the two World Wars. Their number has continued to grow to the present day, pesticides at present numbering in the hundreds, and possibly in the thousands.

Some Problems with Pesticides

The most dramatic changes in the methods of farming occurred during the latter half of the twentieth century especially with the introduction of monoculture ie large areas where only a single crop was grown. Whereas in the past, the system of crop rotation was a useful way of controlling some insect pests, the modern monoculture system is believed to have increased the susceptibility of crops to insects and diseases and, concomitantly, our reliance on the use of pesticides.

While there is little doubt that monoculture, in combination with the use of pesticides and mechanisation, has played a major role in greatly increasing our food supply, many now believe that this success has been achieved at some cost. Resistance to many of the pesticides is a problem. A further problem, obvious in hindsight, is the non-discriminatory nature of many of the pesticides. This has resulted in the killing of the natural predators of insect pests thereby further increasing reliance on chemical control methods. Yet another problem relating to the indiscriminate use of pesticides is their possible effect on the delicate balance of the environment. Apart from their non selective nature, some, notably the organochlorine insecticides, are broken down very slowly and therefore accumulate in the environment. Finally, there have been increasing concerns over the possible impact of pesticides on human health.

These concerns have prompted the development of more selective, biodegradable pesticides that can be used in much smaller concentrations to achieve similar results but without the same environmental impacts. Despite this, concerns about the environmental and health effects of pesticides remain.

How Do Pesticides Work?

The non chlorination pesticides are broadly divided into three separate categories – insecticides, herbicides and fungicides – for the control of insects, weeds, and fungi respectively. Pesticides are mainly synthetic ie they are produced chemically. However, some pesticides, for example the biopesticides, are mostly derived from living organisms, particularly bacteria [2]. Small amounts of antibiotics are used for bacterial diseases infecting plants but these will be discussed later. Viruses also attack plants but non chemical methods are normally used to control infections.

Insecticides

Insecticides are used at all stages in the growing of food – from the fumigation of soil prior to seeding, to application at different times of the growth cycle, and even during storage. There are a number of classes of insecticides, each with their own specific effects on insect pests.

The major classes of insecticides include the organophosphorus, organochlorine, organosulfur, spirazoles, spinosyns, formamidenes, carbamates, pyrethrin, and neonicotinoid insecticides [3]. In addition, the last few years have seen the development of a completely different class of insecticides - protein extracts of certain bacterial strains (Bacillus thuringiensis in particular). In another development, and not without some controversy, through genetic engineering, genes for these same insecticidal proteins have been inserted into certain crops (eg canola, corn), the result of this being that the plant itself generates its own insecticides and therefore may not require insecticide sprays at all. Of course, genetic engineering of plants to produce foreign proteins is not without controversy but will not be discussed in this book.

The organophosphorus insecticides, as the name implies, are organic substances ie they contain carbon atoms, and an atom of phosphorus. They include the well known dimethoate, malathion, and mevinphos. The organophosphorus insecticides work by blocking the effect of a substance, acetylcholine, that is critical for the nerve to nerve, and nerve to muscle transmission of impulses. Acetylcholine is not unique to insects because it is also found in human tissues and is similarly involved in the transmission of nerve impulses. Not surprisingly, therefore, many of the organophosphorus insecticides are highly toxic to humans. In fact, the development of organophosphorus insecticides stemmed from work carried out during the Second World War on related substances, the nerve gases tabun and sarin. The toxicity of organophosphorus insecticides prompted the development of the neonicotinoids, a class of pesticides that also interfere with the action of acetylcholine but are less toxic.

The organochlorine insecticides include the now notorious DDT, aldrin, dieldrin, endosulfan, and chlordane. As the name implies, they all contain at least one or more chlorine atoms. The organochlorine insecticides are also toxic to the nervous system of insects where they are believed to block the movement of metals (or strictly speaking positively charged metal particles termed cations) in and out of nerve cells. Because they are often more soluble in fat than water, they can accumulate in fatty parts of human and animal tissues. Many of the organochlorine insecticides are relatively stable and can accumulate in soil and water for long periods of time. Because of this relative stability, and their predilection for fatty deposits of tissues, they can build up to very high levels in the tissues of some animals.

It is well known that the tissues of certain plants contain substances which act as insecticides and these extracts are true biopesticides. These so-called "botanicals" include neem, derris dust, citrus oil, and pyrethrum. The active ingredients of pyrethrum are pyrethrins and these plants may contain six or more different pyre-thrins. Identification of the chemical structures of these naturally occurring substances led eventually to the generation of synthetic pyretherins, some of the best known being permethrin, cypermethrin, deltamethrin, and fenvalerate. It is believed that the pyrethrins achieve their effect mainly through their ability to inhibit the generation of nerve impulses in insects.

There is yet another way of controlling insects and that is via the use of pheromones. Pheromones are chemicals that attract insects and affect their behaviour. They are not normally toxic to insects or humans.

Herbicides

Herbicides are used to increase the yield of crops by controlling weeds that compete with crops for nutrients, light, and moisture. They may be used to fumigate the soil before sowing, after sowing before any crop emerges, and after emergence of the crop. Depending on the herbicide, the effect is produced either by direct application to the foliage, or by uptake by the roots of weeds. Many herbicides are able to act selectively on weeds without having a significant effect on agricultural crops, while others are non-selective with the potential to kill both crop and weed. The use of selective herbicides relies on an understanding of biochemical and structural differences that may exist between crops and competing weeds.

The range of herbicides now available is staggering. Some of the main classes of herbicides (each class has its own basic chemical structure) are shown below. Examples of some individual herbicides belonging to each class are shown in brackets.

- quaternary ammonium (paraquat and diperquat)
- organophosphorus (glyphosate)
- phenoxyacetic acid (2, 4, 5 T, mecoprop and dichlorprop)
- nitrile (bromoxynil and ioxynil)
- triazine (simazine and atrazine)
- sulphonurea (chlorsulphuron)

Like the insecticides, almost all herbicides are chemically synthesised. Recently, a new class of herbicides, so-called "bioherbicides" which are not chemically synthesised but extracted from certain plants have been discovered [4]. Bioherbicides produce their effects in many different ways and it is likely that their numbers will increase over time. In addition to chemicals extracted from plants, certain bacteria, fungi and even viruses have been used to control weeds [55].

Fungicides

Fungicides are used to control infection of crops at different stages in the growth cycle of plants. The source of fungal infection may be in the soil, on the seeds, or airborne. Some fungi can be controlled by treating the seeds or the soil with a fungicide, while others are controlled by treating the plant itself. Some of the best known fungal infections include those that cause downy mildew and potato blight. These fungi require water for infection of the plant and hence infection is favoured by damp conditions. Other fungal infections, for example those that cause powdery mildew and rust, are not dependent on water for infection. The range of chemicals used to treat fungal infections is also vast. In addition to the chemical fungicides, biofungicides, which are mostly naturally occurring bacteria or fungi, are being increasingly investigated as a means of controlling fungal pests [56]. Some of the main classes of chemical fungicides (with specific examples in each class in brackets) are shown below.

- Copper (Bordeaux mixture)
- Organosulphur (Thiram, Zineb, Captan)
- Organochlorine (Dicloran, Chlorothalonil)
- Dithiocarbonates (Ziram, Ferbam)
- Benzimidazoles (Benomyl)

Different fungicides work in different ways. Some sequester trace metals that are essential to the growth of the fungus. Others inhibit the action of fungal proteins involved in a range of cellular activities such as the generation of energy and the synthesis of fungal hormones.

How Poisonous Are Pesticides to Humans?

In order to understand better the nature of poisons, and many pesticides certainly fit into this category, it is important to note that many chemical substances can produce different toxic effects. A single dose of a poison that has an acute effect may be sufficient to cause death, and a good example of this is cyanide. Amounts of some poisons that are insufficient to kill may, nevertheless, cause serious irreversible damage to a part or parts of the body. For example, some of the toxins in toadstools can cause severe liver damage, some strains of the bacterium E. coli produce toxins that can cause lasting kidney damage, while carbon monoxide poisoning may cause irreversible brain damage. On the other hand, when an animal is exposed to small but repeated doses of some toxic substances for a period of time (chronic exposure), there may be no obvious effect on the animal even though closer examination may show an abnormality in some part of the body.

How Is Toxicity Resulting from Chronic Exposure Detected?

In order to detect abnormalities that may have developed from chronic exposure (ie exposure to small amounts over a long period of time) to a particular substance, some sort of internal examination or some other procedure such as radiology, biopsy, or ultrasound is required. Providing the changes are relatively minor, or if only a small part of the organ is affected, there may be no overt symptoms. Without an internal examination, it would appear that the substance has no toxic effect. A good example of these sorts of changes is the effect of cigarette smoking on the lung. On the other hand, if the exposure continues for a longer period of time, these changes may become more pronounced, or a greater part of a particular organ may be affected until finally the function of the organ is affected and symptoms begin to develop. Once again, cigarette smoking is a good examplewith the development of emphysema and lung cancer. Other examples include the loss of lung function due to exposure to silica (silicosis), coal dust, or even asbestos (mesothelioma).

How Do We Measure the Toxicity of Pesticides?

Examining the acute effects of a pesticide is relatively straight forward and normally carried out with animals, mainly rats and mice. The measure of acute toxicity in an animal is termed the LD50, which is the amount of the poison, expressed in milligrams per kg of body weight of the test animal, required to kill half a randomly chosen group of animals. Because the toxicity of a chemical substance can depend on other factors, measurements also take into account the sex and age of the animal, and pregnancy, and possible teratogenic (ie the capacity to cause birth defects) effects. Another measurement used to assess acute toxicity is the LC50. It is the concentration of a chemical substance in water or air necessary to kill 50% of a group of animals.

Over the years there have been increasing criticisms of the LD50 test culminating . in the anouncement by the Organisation for Economic Cooperation and Development (OECD) in December 2002 that the test would be deleted from its manual of internationally accepted chemical test guidelines. Apart from the quite genuine concerns about the severe suffering of animals used to perform the tests, and the large number required, the test has been criticised for being unreliable and producing inaccurate results. Three alternative procesures – the fixed dose test, the up and down procedure, and acute toxic class, all using fewer animals, are being increasingly used to limit animal suffering. Despite this, the LD50 test has formed the basis for the measurement of the toxicity of most of the pesticides currently in use.

Assessment of Toxicity of Pesticides in Humans

The expression of the toxicity relative to body weight is necessary because of the need to extrapolate the LD50 values to humans. The test is really directed towards determining the likely toxicity of a substance to humans and the LD50 values, even though they are based on data obtained from other animal species, do provide an indication of the likely toxicity in humans. However, because there may be variations in the relative toxicity of a particular poison between animal species, toxicity tests are normally carried out on two or more species. The LD50 values for different pesticides vary greatly with the lower the value the greater the toxicity. LD50 values (using adult male rats) for insecticides range from 2 for phorate to 5000 for methoxychlor, for fungicides the range is around 30 for ethylmercury to 9000 for captan, while for herbicides the figures vary from about 50 for dinoseb to 6600 for dalapon. The extremely wide range indicates that, for rats at least, there are very large differences in the toxicity of the pesticides. Based on LD50 values, the insecticides are considerably more toxic than herbicides and fungicides.

While it is possible to get an indication of the toxicity of most of the pesticides, the chronic effects that result from repeated exposure over extended periods of time, perhaps a decade or more, are much more difficult to evaluate. Studies to determine chronic effects, within the context of the very long term exposure periods of humans, are clearly not practical. However, there have been a number of studies on the effects of repeated doses of very small amounts of pesticides, particularly those such as the organochlorines that are fat soluble and may therefore accumulate in body fat over time.

The results of these studies indicate that repeated doses do not necessarily lead to a steady increase in their levels in body tissues. It is believed that the reason for this is that a balance eventually develops between the uptake and excretion of the pesticides culminating eventually in amounts in tissues that may not change greatly over time. It has also been suggested that the pesticides may stimulate an increase in detoxifying mechanisms within the liver and this may help keep pesticide levels fairly low.

Assessment of potential chronic toxicity is determined by administering varying amounts of a particular pesticide to a number of animals of a particular species and determining the maximum concentration at which no adverse effect on any of the animals studied has occurred. This dose is known as the No-observed Adverse Effect Level (NOAEL) and differs considerably from the LD50. To determine whether any effect has occurred, any weight change is recorded, and blood, urine and faeces are analysed. The animals are then killed and both a macroscopic (ie examination by eye) and microscopic examination of body parts are undertaken. As the measurement of the NOAEL is made in animals, mostly rodents, this figure is divided by what is termed a safety factor, which may vary from 10 to 1000 or more, to convert the NOAEL to the acceptable daily intake (ADI) for humans. The safety factor allows for the differing susceptibilities of the animals and humans to a particular pesticide, and the fact that some humans may be much more susceptible to a particular pesticide than others. It is believed that the use of a safety factor greatly reduces the risk of harm caused by any unforseen chronic effects of pesticides.

Regulation of Pesticide Use

Although the ADIs for different pesticides are based on toxicological data obtained from animals, and incorporate the use of safety factors that are selected, in the main, arbitrarily, they are considered of value by international organisations and individual governments, and form the basis for tolerance limits for pesticide residues in food and water. The United Nations set up the Codex Alimentarius Commission, an international organisation with representatives from many countries, to look into food quality. The Codex Committee on Pesticide Residues was established because of concerns about the presence of pesticide residues in food and their potentially harmful effects. One of its main functions was to define the acceptable limits for pesticide residues. The committee established Maximum Residue Limits (MRLs) for individual pesticides based on the ADIs but taking into account other factors as well. In addition to its role in establishing MRLs, the committee also provides recommendations on the registration, storage, use and analysis of pesticides.

Individual governments throughout the world are guided by recommendations of the Codex Committee, and have set up their own national bodies to oversee the use of pesticides at a national level. The Environmental Protection Agency (EPA) is the government body that oversees the registration of pesticide usage in the USA, while the Chemicals Registration Directorate (CRD) of the Health and Safety Directorate regulates pesticide usage in the United Kingdom. The Australian Pesticides and Veterinary Medicines Authority (APVMA) formerly the National Registration Authority (NRA) in Australia is responsible for administering the use of pesticides. There are similarities in the different regulatory systems but, for convenience, we will focus on how the Australian Government regulates pesticide usage.

Before any pesticide is introduced into the market place, the government agency overseeing the registration process has to be satisfied that, if used in accordance with the directions on the label, the product is not a danger to human health or to the environment. An assessment of the potential hazards of the pesticide is made based on the LD50, ADI and MRL values. Approval is only given for a particular purpose. Thus if a particular pesticide is approved as a spray for apples, it cannot normally be used, for example, for cherries or zucchini without further supporting data. The APVMA has to be satisfied that a person eating the fruit or vegetable is not exposed to pesticide levels in excess of the MRL. Therefore, pesticide manufacturers have to indicate the levels of pesticide residues remaining if a user followed the instructions for use provided at the time of sale.

In the US, the licensing and monitoring of pesticides is carried out by the Environmental Protection Agency. Applicants wanting to register a pesticide for use have to provide details on the ingredients of the pesticide, the type of crop that is to be treated, the frequency and timing of use, and storage and disposal practices.

The EPA assesses the likely environmental effects, and effects on human health from information provided by the applicant who, in turn, have to adhere to guidelines for testing laid down by the EPA. http://www.epa.gov/pesticides/factsheets/registration. htm

Monitoring of Pesticide Levels in Food

A number of US government agencies are involved in the monitoring of pesticide residues in food, including the Federal Drug Adminstration Bureau (FDA), the USDA (US Department of Agriculture), and the EPA. The latter is involved in setting the maximum residue levels in US grown and imported fruits and vegetables. In Australia, the APVMA has to be satisfied that individual pesticides are safe to use, but it is not involved in the monitoring of pesticide levels in food. In Australia, there is little coordination at a national level with various organisations taking the responsibility for monitoring. There is a National Residue Survey, a program conducted by the Department of Agriculture, Fisheries and Forestry Australia and funded mainly by industry, which carries out random checks on pesticide levels, and the Australian Total Diet Survey carried out by Food Standards Australia New Zealand. Some corporations measure pesticide residues in some of the food they sell and there is an Australian Milk Residue Analysis Survey (coordinated by Dairy Food Safety Victoria on behalf of the Australian Dairy Authorities Standards Committee and the dairy industry) which looks at agricultural and veterinary chemicals in raw cow's milk In South Australia, the State Government, the Department of Human Services (formerly the Health Commission) and the wholesale fruit market at Pooraka (referred to as the Pooraka Food-Care Project) carry out random checks on pesticide residues in market produce and other states operate similar schemes.

Recommended levels of pesticides in food in Australia are determined by Food Standards Australia New Zealand. ADIs for most of the pesticides are mostly in the millionths of a gram range per kilogram body weight, while MRLs ie the maximum permitted residues in food are in parts per million or even parts per billion.

What Sorts of Levels of Pesticides Are Present in Our Food?

What we know about pesticide levels in food is clearly only as good as national and state government capacities to monitor residue levels. In developed countries, for example the US, Europe, and Australia, there is relatively close monitoring of pesticide levels to ensure that levels are kept well below the MRL.

It is worth looking at what one of the agencies, in this case the US FDA, found for the year 2008 [5]. The FDA monitored residues in both US produced food as well as food imported from 93 countries. They looked at over 5000 samples from US and imported food and found no residues in 64% of US food and 72% of imported food.

Of course this means that pesticides were present in around a third of all US food, and a quarter of all imports. Less than 1% of US food had residue levels in excess of the guidelines established by the EPA, but almost 5% ie 1 in 20 of the imported foods had levels in excess of these guidelines. Eighty three pesticides were found in the different foods, and another twenty five degradation products, contaminants, and related substances. Similar results were observed with both domestic and imported animal feeds. A more recent analysis (2013) has detected 219 separate pesticide residues in different foods [57].

It is of interest to look at the data reported by the FDA in some detail because it provides information on the extent of pesticide contamination our food. For example, analyses of whole wheat bread that is an important component of the diet of many Americans, has confirmed that, at least at the time the analyses were carried out, twelve pesticides were detected. Assuming 100 grams of whole wheat bread bread consumed per day was 100 grams -which is around about the size of a bread roll –the total combined amount of pesticides consumed in bread alone could be as high as 20 millionths of a gram. Seven pesticides, averaging nearly 7 micrograms in total per 100 g, were found in pork sausage and 60 micrograms per 100 grams of 15 pesticides in peaches. Obviously, the amounts and numbers of pesticides in, for example, a peach vary considerably and almost certainly some will have more and others less but it gives us a general idea of what is in our food. What these data tell us is that there are a significant number of pesticide residues in food and that while the amounts are small, they are significant.

Just in case anyone thinks that this is a peculiarly American problem, we can look at some of the Australian figures. One of the more recent analyses of the total dietary intake of contaminants and nutrients was carried out by Food Standards Australian and New Zealand as part of the 23rd Total Diet Study (ATDS) and results published in late 2111 [6]. The focus here was on the estimation of the likely intake of pesticides in a typical Australian diet. The intake of each pesticide is expressed as a percentage of its ADI. Levels of 214 agricultural and veterinary chemicals and pesticides were measured in 92 different foods and drinks and the likely daily dietary intake of these chemicals was determined for a number of different age groups. Pesticides were found in most foods – meat, milk, fruit, grains, and vegetables. The conclusions were similar to those reported by the FDA ie the intake of most individual pesticides was not very great, mostly averaging less than one microgam per day, and considerably less than 1% of the ADI. However, some, notably the fungicides dithiocarbamate and iprodione, and the insecticides propargite, chlorpyrifos, and piperonyl butoxide, were present in much larger amounts than this. Moreover, the ADIs of these pesticides were greater than 1%, in some cases approaching 60%. It is worth noting that, because of the large numbers of pesticides found in the many different foods, the total amount of pesticides consumed was significant (more than 100 micrograms daily) even though the levels of individual pesticides were small. It is a little like a grain of sand – not much there but, if you have lots of grains then you have a whole beach. Residues from literally dozens of pesticide residues were detected in a great number of foods, including meats, vegetables, fruit.

Based on these figures, an average 70 kg adult males' total daily consumption of pesticides could be in excess of 150 micrograms, particularly if diphenylamine, a chemical used to prevent the deterioration of fruit such as pears and apples, is included. What is even more worrying is that the pesticide intake of children under five years on a body weight basis is around double or more the intake of adults.

Is Government Regulation of Pesticide Use Adequate to Protect Us from Harm?

This clearly is the $64,000 question. Registration of new pesticides is an expensive, tedious, and time consuming process and certainly not for the financially challenged. As discussed earlier, before permission is given, a particular pesticide has to be shown to be effective for its intended use, safe for people and animals, and not pose an environmental threat. Toxicology data have to be assessed by the appropriate government agency with input from other agencies (eg EPA, FDA, USDA in the US and APVMA, Therapeutic Drug Administration Bureau (whose main role is to vet potential drugs and other therapeutic agents), the Department of Environment and Heritage (which evaluates potential environmental hazards), the National Occupational Health and Safety Commission (which oversees the likely impact on people exposed to the chemical in the work environment eg farmers, manufactures), Food Standards Australia New Zealand (which is responsible for ensuring the safety of food), and the Australian Quarantine and Inspection Service (whose role relates to ensuring the safety of imported and exported food in Australia. It is important to stress that the assessment process is totally dependent on the information provided to the government agency by the applicant as, these agencies do not normally do any testing themselves but instead, they examine the data provided by the applicant and decide whether a particular pesticide has the capacity for harm in the amounts likely to be present in food. They may use their own staff or approach others whom, they believe, have the appropriate expertise to properly assess the information they have been provided with. In certain situations they may even seek assistance from overseas bodies.

Some appreciation of the complexity of the process can be gained by inspection of the registration section of the EPA and APVMA websites. Before permission is given for use of a new pesticide, a large amount of experimental information on toxicology measurements in at least two different animal species, has to be provided by the applicant. Because of its highly specialised nature, the toxicology is mostly contracted out by the applicant to companies that specialise in this type of work. One advantage of this for consumers is the fact that it reduces the possibility of applicant bias in the studies. However, the possibility of bias cannot be totally excluded as the applicant funds the studies and also provides the samples to be tested.

The approval given by the government agency relates to the specific use of a particular pesticide. As part of this approval, the applicant has to demonstrate that the pesticide residues on crops after treatment are below the MRL. However, before, for example, the APVMA-approved pesticide can be used on food crops, further approval is required from Food Standards Australia New Zealand who have to be satisfied that the MRL is realistic given the likely consumption of the pesticide in the overall diet.

Superficially, the registration scheme would seem to be very effective. However, as discussed above, it is very much dependent on information provided by the manufacturer or company that is anxious to have its product endorsed, and by the assessment of that information by staff from a number of government bodies. As in any other large government bureaucracies, it is possible that decisions made by staff as part of the assessment process could be affected by work load and perhaps even political pressure. And, while the MRL for a pesticide may be appropriate at the time the assessment is made, further research may indicate that a much lower value is warranted. One example of this is the relatively recent report that MRL values for a range of pesticides, particularly those that are toxic to the brain, may be set too high for infants because of their potentially greater susceptibility to this class of pesticides [17]. Then there are other examples where a pesticide has to be withdrawn from sale because further research has shown that there are hazards which were unsuspected at the time of registration eg endosulfan and aldrin [58, 59].

The monitoring process is also not without its flaws. While there is no doubt that monitoring is carried out on a regular basis by government agencies, and sometimes by industry, it is undeniable that only a very small proportion of food or crops are subjected to any form of chemical scrutiny. The process is completely random and depends very much on the results obtained from these random samples accurately reflecting the overall picture. It is not difficult to see how easy it would be to abuse the system as the chances of being caught do not appear to be that great. Indeed small numbers of violations (ie pesticides for which there is no permit for use on a particular crop) are detected even in monitored samples which makes one wonder how widespread the practice really is in unmonitored samples. The other potential problem with monitoring is that, for practical reasons, it is very difficult to look for all of the very large number of pesticides now available.

And finally, the analytical procedures that are used are unlikely to detect any contaminants unless they are closely related chemically to the pesticides being monitored. Some of these may be even more toxic than the pesticide itself. Contaminants may be introduced in different ways. They may be introduced during the manufacturing process, during storage, or they may be present in the supposedly non-toxic vehicle used to suspend or dissolve the pesticide. One illustration of the devastation caused by contaminants relates to the problems experienced a number of years ago by orchid growers. In this case, a particular pesticide routinely used to control fungal infections in orchids as well as in other crops, caused wide scale destruction and severe economic loss to growers. The damage to crops seemed to be caused only by a particular batch of the fungicide indicating the presence of a contaminant. While in this particular situation, the public was alerted to the presence of a toxic contaminant by the destruction of crops, it seems possible that there may be other contaminants that do not cause such obvious effects on crops and are therefore likely to be missed. If these contaminants are toxic to humans, there is a real concern if they enter the food chain.

Summary – Pesticides (General)

• There is a great variety of insecticides, herbicides, and fungicides used at all stages in the growing of food.
• Their use has contributed to the quality and choice of food available as well as the cost.
• There are environmental and ecological concerns in the indiscriminate use of pesticides.
• Testing of the toxicity of pesticides involves their administration to animals, mainly rodents, and determination of the levels required to produce effects in animals.
• The amounts of individual pesticides detected in foods by random monitoring carried out by state governments and industry rarely approach the maximun residue level (MRL), an amount considered to pose no threat to human health.
• The MRLs of individual pesticides permitted in food are a thousandth or less than the amounts producing a measurable effect in animals.

What Is the Evidence for Toxicity of Pesticides in Humans?

Given that most of the toxicity studies on pesticides are carried out in rats and other animals, what evidence is there they are toxic to humans? To examine the evidence for toxic effects, we really need to look at both the acute and chronic effects.

Acute Toxicity   AU!

There is a considerable body of evidence in the medical and scientific literature that exposure of human beings to pesticides can cause death or serious injury. Many of the reports on the acute toxicity of pesticides have related to suicide attempts but these reports are of interest because they provide an indication of the parts of the body that are affected. Acute poisoning with organophosphorus insecticides can result in death [7]. The major affect of these substances is on nerve and brain function. Post mortem analysis of tissues has confirmed that there is build up of the chemicals in many different organs, not just the brain. Even exposure to the newer, less toxic, neonicotinoid insecticides have been reported to cause harm and death [8].

Deaths have also been reported in individuals from some of the more toxic herbicides, for example diquat and paraquat [9]. Severe ulceration of the intestine, and kidney and brain damage, are commonly observed. Even some of the less toxic herbicides can cause death. For example, attempted suicide by taking an overdose of glyphosate mixed with a surfactant (a substance used to increase the "wetting" of the foliage of plants) is not that uncommon. Glyphosate is one of the most commonly used herbicides. In a review on the effects of glyphosate intoxication, it was reported that as many as 8% died after taking the herbicide, and lung and kidney damage were commonly observed [10]. The toxicity of glyphosate is of considerable interest bearing in mind its apparently very low toxicity in rats. Indeed, in a Fact Sheet produced by the SA Department of Primary Industries (FS 1/96) [11], it is stated that glyphosate is less toxic than common salt. The dose required to kill 50% of a group of men and women weighing around 70 kg would be around 350 g. As much of the glyphosate sold is at a concentration of 100 g per litre, the average person would need to take about 3500 ml, a staggering amount and one that it is difficult to imagine anyone taking. It makes one wonder whether the herbicide is more toxic in humans than it is in rats, or whether the surfactants and other components added to the herbicide are even more toxic than the herbicides. In smaller, but non-fatal amounts, pesticides have even been reported to mimic some of the symptoms of food poisoning. Thus an outbreak of what was thought initially to be a food-borne illness in Taiwan with dizziness, headaches, nausea, weakness, and vomiting, was later shown to be caused by contamination with the insecticide methomyl [12].

Chronic Effects

Any assessment of the chronic effects of pesticides on humans should include an examination of whether they may increase the risk of degenerative diseases such as cancer, and whether exposure increases the risk of birth defects. Chronic effects are those resulting from repeated small doses, in the amounts likely to be present in a normal diet and therefore insufficient to kill or cause serious harm. It is important to reiterate that, while animal studies are important in assessing a possible risk of pesticides in humans, the most relevant studies would be those conducted with humans. However, because of the toxicity of many of the pesticides, it is unethical to deliberately subject human subjects to repeated doses of known amounts of these substances over the long term.

The sorts of questions we would like to have answered include the following. Does repeated exposure to very small amounts of pesticides cause disease? If they do cause disease, what sort of exposure is required and for how long? Are there levels of exposure below which there is no increased risk of disease and what are those levels for the various pesticides? Are all pesticides the same or are some more likely to produce disease? Are some humans more susceptible to harm on exposure to pesticides and, if so, why is that? These are very difficult questions to answer even with animals whose diets and environment we are able to control more effectively than we can with humans.

As we cannot resort to human experimentation, we have to rely on other methods. The only practical way we can look at the effects of pesticide exposure is through population studies. These are called epidemiological studies and involve the collection of data from large groups of individuals who may have had different exposures to individual pesticides. For example, a comparison of the health of two different groups with markedly different exposures over long periods can provide an indication of any negative effects. There are different groups in the community who, because of the nature of their occupation, have had a higher than normal exposure to pesticides. These include farmers, horticulturists, silo workers, pest exterminators, crop dusters, factory workers, and plant nursery workers. Careful monitoring of the health of these groups can tell us whether a repeated exposure can give rise to any health problems or lead to an increased risk of disease.

Diseases Caused by Occupational Exposure to Pesticides

The answer to the question 'does repeated exposure to very small amounts of pesticides causisease?' appears to be yes. In fact there have been numerous studies conducted over the years which point to the negative impact of pesticides on human health. Many of these studies have involved farmers or agricultural workers and diseases specifically looked for include various forms of cancer, respiratory difficulties, skin problems, brain abnormalities, and blindness. Let us look at a few examples taken from the large number of publications over the last decade or so.

An increased risk of respiratory problems, and eye and skin irritation has been observed in people who had a greater than normal exposure to fumigants [13]. The fumigants included methyl bromide, aluminium phosphide, and metamsodium, and were used to sterilise fields, barns, and grain storage areas. In some instances, the exposure was believed to be due to drift from adjoining fields, in others it was thought that the workers had prematurely re-entered a barn or grain storage area after fumigation. The respiratory problems and eye and skin irritation observed by this group of researchers was confirmation of a report published a year earlier of a study that had been carried out in the United Arab Emirates [14]. The list of symptoms described was very extensive and included bronchitis, pneumonia, asthma, sore throat, shortness of breath, sinusitis, etc. Even workers exposed to fumigants used in shipping containers have also been shown to be affected [60]. The link between occupational exposure to pesticides and and a variety of respiratory problems, including asthma, has been confirmed more recently [15].

Brain Diseases

Parkinson's disease is a brain disease affecting movement. A report linking Parkinson's disease, with pesticide exposure was published in 2000 [16]. In this particular study, the health of a group of more than two million men and women living in Denmark were monitored for 12 years. It was found that those men and women who had worked in agriculture and horticulture, and therefore had been exposed to higher than normal levels of pesticides, had a greater than normal risk of getting Parkinson's disease. In another study published in the same year in the prestigious medical journal, Lancet, a link between diminished mental capacity and pesticide exposure was suggested [17]. These conclusions were based on an analysis of data obtained from the Maastricht Aging Study, a study in which the health of a very large group of individuals was monitored over a number of years. While neither of these two studies has demonstrated a definite link between deterioration of brain function and pesticide exposure, they are clearly worrying and require further investigation. More recent reports have confirmed this apparent link between pesticide exposure and increased risk of Parkinson's disease [18, 61]. There are indications as well that exposure to environmental factors, including pesticides, may increase the risk of another degenerative disease affecting the brain, motor neurone disease or amyotrophic lateral sclerosis [19, 62].

Birth Defects

There are reports linking birth defects to pesticide exposure. One group of researchers observed birth defects in four children born to mothers who had been inadvertently exposed to the insecticide chlorpyrifoas [20]. The defects included abnormalities to the brain, eyes, ears, palate, and teeth. These findings were later disputed by the US Environmental Protection Agency. In a study carried out in the Eastern Cape Province of South Africa it was found that there was a seven fold increase in birth defects in women exposed to pesticides through their activities in gardens and fields [21]. More recently, it was suggested that birth defects in two children (heart, anus, absence of growth hormone) may have been due to exposure of the parents to pesticides [22]. Exposure to pesticides has also been reported to increase the risk of neural tube defects, an abnormality of the spinal cord [63]. There seems little doubt that pesticides are taken up by the foetus from the mother during pregnancy [23].

Cancer

Authoritative government agencies such as the US EPA, and non government bodies such as the WHO and the Interantional Agency for Research into Cancer (IARC) have stated that some pesticides are carcinogenic and, at the very least, can cause cancer in animals [63, 64]. However, it is not known whether pesticides do cause cancer in humans. There are suggestions that there may be a link between exposure to pesticides in the workplace and cancer but a definite relalationship has not been established. Part of the reason for this is that cancer is not a single disease but is really a group of diseases, each disease primarily affecting one of many parts of the body. Despite this, however, there is enough information to raise concerns. A study carried out in Brazil appeared to show that agricultural workers have a higher risk of developing cancer, specifically cancers of the oesophagus, stomach, and larynx and suggested a possible involvement of pesticides [24]. Pesticide exposure has also been linked to an increased risk of lung cancer [25].

There have been suggestions that pesticide exposure may increase the risk of cancer of the pancreas [26]. One of the major functions of the pancreas is the production of insulin, a hormone that regulates blood sugar levels. The types of pesticides that have been thought to increase the risk of pancreatic cancer are the organochlorines A more recent study has supported the apparent association between pancreatic cancer and pesticides although other research has disputed this finding [27, 28].

The possible involvement of pesticides in the development of cancer of the prostate was suggested by a large study carried out in Florida and published in 1999 [29]. The study looked at the health of more than 33, 000 licensed users of pesticides in the 19 year period from January 1975 to December 1993. While the general health of the pesticide users was assessed as being better than the general population of Florida, the study showed a statistically significant increased risk of prostate cancer in licensed pesticide users. These findings have been supported by more recent research [30].

There have been reports in the medical literature linking pesticide exposure to certain types of childhood cancers. A German group of researchers investigated the relationship between pesticide exposure and cancer in more than 2000 children with leukemia, non-Hodgkin's lymphoma and various other cancers [32]. They observed a strong relationship between the cancer and the degree of parental exposure to pesticides. The risk of cancer increased in response to an increased household exposure either through parental or professional use of pesticides to exterminate pests. There was also an increased risk of leukemia in children living on farms. Research carried out in the US also suggests that there is an increased risk of lymphoma in the children of parents exposed to pesticides [31].

There are two conclusions to be drawn from the various reports linking pesticides to cancer. Firstly, they do not prove, beyond reasonable doubt, that pesticides cause cancer. When we deal with human populations, there are too many variables. What the studies do is indicate that there is an association of some sort between cancer and pesticides. Almost certainly, there are a number of other factors, for example genetics, diet, sex, age, general health etc. that, together with pesticides, influence the risk of getting cancer. Unfortunately, at this stage, it is not possible to predict who is more likely to get cancer because of pesticide exposure even though we can say that there may be an increased risk. Secondly, we know very little about the level of exposure required to increase the risk of cancer, nor do we know which of the chemicals used are the most potent cancer-producing agents. Further studies are necessary for us to be able to answer these questions.

Known Unknowns?

More recently there have been reports that some pesticides which have been shown to be relatively non toxic, at least in animal studies, may have more subtle effects that were not forseen. The best example of this is the herbicide glyphosate. There are indications that glyphosate can act as an antibiotic.with possible effects on beneficial gut bacteria [33]. It has also been suggested that glyphosate residues in well water may be a factor in the development of kidney disease in Sri Lanka although the evidence for this is not definitive [34].

Diseases Caused by Exposure to Pesticides Outside the Work Environment

The most difficult question of all, ie whether there is a risk of disease in people who are not exposed to pesticides in their workplace, has not been adequately addressed. There are potentially many different methods of pesticide exposure -through atmospheric drift to properties adjoining farms, drift that occurs after aerial crop dusting, or household or garden exposure after spraying with pesticide sprays, to list just a few. For most people, the principal means of exposure is through the food they eat [35]. However, there are very few studies that have demonstrated any link between pesticides in food, in people who live in the developed countries, and disease. This is not because there is no such relationship but rather because the right studies have not been carried out. The types of studies required to show such a relationship are complex and require a comparison between a population eating almost exclusively food grown in the complete absence of pesticides, and another group enjoying conventional food. These studies are complicated because the type of diet can contribute to the development of degenerative diseases [36], so the two populations being compared should have similar diets apart from the absence of pesticides in the food of one of the groups.

One of the very few such studies involved a comparison of the semen quality of Danish farmers who were divided into three groups according to their intake of organic food ie food grown in the absence of pesticides and synthetic fertilisers [37]. Semen samples were obtained from members of each group and a number of characteristics, such as sperm count, numbers of dead spermatozoa, degree of movement, numbers of normal spermatozoa, and percentage of normal sperm heads, were analysed to determine whether there were any differences in the quality of the spermatozoa which could be related to the intake of pesticides. While the number of organic farmers studied was not very large, nevertheless the data indicated that there were very few differences in sperm quality between the three groups apart from the lower number of spermatozoa considered to be normal in the group with the highest intake of pesticides. The authors concluded that the levels of pesticides in the diet did not appear to increase the risk of a lower quality semen but they cautioned that this conclusion could not be applied to the general population because the degree of dietary pesticide exposure, and the types of pesticides consumed probably varies significantly from individual to individual.

There is no doubt that consumption of organic food, with its lower levels of pesticides, does result in a reduced exposure to these substances, providing of course there is no exposure through occupation or other means [38]. Further evidence for this can be found by measuring the amounts of pesticides in the urine which is reduced in those consuming an organic diet [38, 39]. Despite this, a comparison of the health of individuals consuming conventional and organic foods has failed to demonstrate any significant difference.even though a very recent study carried out by a group of Norwegian researchers seems to have shown that the risk of pre-eclampsia is lower in pregnant women on an organic diet [40]. Pre-eclampsia is a potentially serious condition in pregnancy and results in high blood pressure, swelling of tissues, and loss of protein in the urine. There is also a suggestion that the risk of developing one particular form of cancer, non-Hodgkin lymphoma, may be slightly reduced [41]. However, there is a suggestion that meat produced by conventional methods may be more likely to contain antiobiotic-resistant bacteria.than organic meats [42].

Given that we all consume small amounts of pesticides, are there genuine grounds for concern? The fact that there are few studies demonstrating a link between dietary pesticides and disease is not necessarily reassuring because such studies are very difficult to carry out. Perhaps the most persuasive argument against any such link is the fact that there are only very low levels of pesticides present in our food, levels far below those required to cause harm in animals. Now, while it is true that the levels are very low, in some cases even lower than the level of detectability using modern methods of chemical analysis, it does not automatically follow that they are harmless in humans. As mentioned above, there are many factors, including the amount and type of pesticide consumed, and the susceptibility of the individual, that influence their potential to cause harm in an individual.

Varying Susceptibility

At the time of writing, there seems to be some evidence that occupational exposure may increase the risk of disease. One factor in determining whether we get sick or not is the amount we are exposed to, and another is the period of exposure. That is, how much and for how long. Perhaps an even more important factor is our innate susceptibility and this is determined by our genetic makeup. The susceptibility of different individuals to pesticides and other poisons almost certainly varies. This can easily be seen in animal studies where the doses required to produce an effect in individual animals vary even though most studies involve the use of genetically similar animals, and factors such as diet and environmental conditions are not all that different. The situation is more complex in humans because of the very significant genetic, dietary, and environmental factors that exist. In addition, there are other factors, such as smoking, alcohol, and drugs that further complicate matters.

Children

The effect of body weight on a person's response to a drug or chemical (generally speaking smaller amounts of a drug are required for smaller people to produce the same effect) is well known, but there are other factors too. Children are believed to be particularly sensitive for a number of reasons [43–45]. The dietary patterns, particularly for very young children (under a year) are very different. Because of their rapid growth, they require between 3–4 times more food on a weight for weight basis than adults, and this food is less varied than that consumed by adults [46]. Consequently, it has been suggested that their consumption of certain food items may be many times greater (again on a weight for weight basis) than adults.

A more potentially worrying source of pesticides for very young children is breast milk. Numerous studies have demonstrated the presence of these substances in milk [47]. In developed countries, the levels are very small, and appear to be influenced by the diet of the mother. However, even though the levels are small on average, there is considerable variability and, in some cases, the amounts present may be in excess of the ADI. It is easy to see, therefore, that children may be more likely to be exposed to not insignificant amounts of certain pesticides.

Also, it is not appropriate to consider very young children as simply mini-adults with similar suscept ibilities to adults. At birth, the various organs, particularly the brain, the immune system, and the reproductive organs are not fully developed and are believed to be potentially more sensitive to environmental poisons than fully developed organs. There are possibly differences in the abilities of very young children, compared with adults, to absorb and excrete toxic substances, and in the eventual distribution of such substances in different parts of the body. Following absorption from the gut, potentially toxic substances travel in the blood to the liver where they are detoxified and released back into the bloodstream where they travel to the kidneys and are then excreted in the urine. However, there is ample evidence that some of these substances can find their way into other organs, particularly the brain and fat deposits throughout the body.

In children, what elicits most concern is the brain. As discussed earlier, many of the insecticides work by blocking a critical process in the nervous tissue of insects, the degradation of a key substance, acetylcholine. Acetylcholine is termed a neurotransmitter, and takes part in the transmission of chemical 'messages' from one nerve cell to another, or from nerve cells to muscle. Acetylcholine is but one of a number of neurotransmitters in nervous tissue that have similar functions. Once the acetylcholine is released from a nerve cell, it is taken up by the adjoining nerve cell or by a muscle cell. If the adjoining cell is a muscle cell, it may respond by initiating a movement such as a contraction or if it is another nerve cell, it may be induced to pass a similar chemical message on to its adjoining cell. However, if the acetylcholine remains attached to the muscle cell, or to the nerve cell, the signals will continue almost indefinitely, so clearly there has to be a mechanism to turn off the signal. This is achieved via a special protein, acetylcholinesterase, found in nerve and muscle, that chemically degrades acetylcholine thereby turning off the signal.

Now acetylcholinesterase is not only found in insects, it is also found in the brains of animals such as rats and mice, and occurs in humans as well. It is known that the degradation of acetylcholine in the human brain is also carried out by a human form of acetylcholinesterase and, as in insects, this is blocked by organophosphate insecticides. However, because of the relatively greater size of humans, much larger amounts of organophosphates are required to produce an effect. There are suggestions as well that organophosphates may be able to affect the brain in other ways.

The brain continues to develop for 18 months or more after birth. Some parts of the brain, for example, those areas that control vision, may continue to develop for 3–4 years after birth. While tests to assess the potential effects of a pesticide on humans, including its effects on brain function, are mostly carried out using rodents, the relevance of these animals in testing for effects on the brain in children has been questioned because the maturity and development processes of the brain in these species differ from humans. During brain development, it is considered that there may be periods when the developing brain is particularly susceptible to environmental contaminants such as pesticides [44]. Animal studies have suggested that exposure to toxins, even for short periods during brain development, can produce abnormalities that are not manifested until later in life. What is particularly alarming is that there may be abnormalities in brain function that cannot be detected by careful examination of the brain at the time of exposure to the chemical. In some instances, the only evidence of an affect on the function of the brain is observed later [48, 52, 53, 54].

Another factor in susceptibility to pesticides is the reduced capacity of an infant brain to exclude toxic substances. The fully developed adult brain is protected, to some degree, by a mechanism that keeps out undesirable chemicals such as pesticides. This mechanism is termed the 'blood brain barrier' and it relies on a restriction of the movement of substances from the blood into the brain. The blood brain barrier is not fully developed at birth in rats, the animals most often used to test pesticide toxicity, but matures at around 3 weeks of age. There are also suggestions that the maturation of the blood brain barrier occurs in different parts of the rat brain at different times. In contrast the blood brain barrier in humans is not believed to be fully mature until a year or so after birth, and little is known of the maturation process in the different regions of the human brain.

Other classes of insecticides, the pyrethrins and organochlorines, have also been reported to be harmful to the developing brain. Administration of pyrethrins to young mice changed their behaviour when they reached maturity even though there was little evidence of any harmful effects at the time of administration of the pesticide. It has been concluded from these experiments that exposure to these two other classes of pesticides in the neonatal period, in amounts insufficient to show any overt evidence of any harmful effects, can also affect brain function later in life [57].

The Unborn Baby

Another group potentially susceptible to pesticide exposure in food is the unborn. However, while there is considerable evidence that alcohol and drugs such as thalidomide can cause birth defects, there is little evidence that pesticides in the materanal diet can cause abnormalities in the developing foetus even though exposure to pesticides in other ways does appear to increase the risk of birth defects [21, 22].

Adults Who Increased Susceptibility

After absorption, most potentially harmful substances travel to the liver in the blood stream. One the most important functions of the liver is detoxification and there are two main steps in this process. Firstly, substances such as pesticides are changed chemically, facilitated by special liver proteins, into products that are much less harmful to the body. The second part of the process is excretion, and this involves a further chemical modification of the pesticides by other liver proteins into a form that can be removed by the kidneys from the blood and then excreted into the urine.

There is plenty of evidence that the efficiency in ridding the body of drugs and other chemicals can vary dramatically in different people [49–51] Some people have been identified as being 'poor metabolisers' of drugs or chemicals ie these processes are not very active, others are termed 'ultrarapid metabolisers' ie these processes are extremely rapid, while some fit somewhere in between. There have even been suggestions that these differences can help to explain the differences in risks of developing cancers that arise as a consequence of exposure to chemicals in the workplace or in the environment. As there is clear evidence from animal studies that pesticides are also changed chemically and excreted by these same processes, it is reasonable to speculate that people may also differ in their abilities to handle these substances and that some may be particularly sensitive. In fact, there is now clear evidence that there are significant differences in our capacities to degrade one of the most lethal classes of pesticides, the organophosphates [51].

Poor liver function, perhaps as a consequence of diseases such as hepatitis and cirrhosis (inflammation and loss of function) of the liver caused by excessive alcohol intake, can also contribute to a reduced capacity to remove pesticides as well as other chemicals. Therefore, people who have impaired liver function, may also have a reduced ability to remove pesticides from their body.

Animal studies indicate that there may be another group of individuals with an increased susceptibility to pesticides in food. As discussed earlier, exposure to pesticides can increase the risk of allergic asthma although the mechanism for this is not known [55]. However, studies with rats have provided a possible clue to a mechanism. Ingestion of certain pesticides has been shown to increase the response to the allergic substances derived from mites, known environmental triggers for asthma in humans [56]. These experiments and others have highlighted the possible role of pesticides in increasing allergic reactions. Thus, people who are prone to allergies may have an increased susceptibility to pesticides through the capacity of these substances to augment their allergic reactions [55].

Studies with animals have suggested another potentially susceptible group. While exposure of very young mice to very low levels of pesticides may have little obvious effect, further exposure in adult years can result in effects on learning, memory, behaviour [54, 57]. These experiments imply that injury to some parts of the brain of the young animals through exposure to the low levels of pesticide had occurred, despite the apparent absence of any obvious signs, and that these particular areas of the brain were more vulnerable to injury from foreign substances such as pesticides, later in life. If we extrapolate these findings in mice to humans, it is tempting to suggest that there may be a group of individuals with a greater susceptibility to pesticides because of their early exposure. Repeated dietary or environmental exposure could therefore increase the risk of a loss of brain function for these people.

Another group of people who may have susceptibility include those with the quaintly named TILT (or Toxicant-Induced Loss of Tolerance) syndrome which is really another name for a chemical intolerance [53]. Multiple chemical sensitivity is a form of this condition. Little is really known about how this condition develops. However, the current view is that the syndrome is a product of two quite separate processes. The first step involves an exposure to any one of a number of chemical substances, for example, those contained in perfumes, cigarette smoke, petrol, pesticide sprays, new carpets, nail polish remover etc. This exposure is thought to break down the body's natural tolerance to chemicals. At some later stage, the second step involving a further exposure to similar chemicals seems to trigger TILT and affects one or more parts of the body including the heart, brain, intestine, muscles, and lungs. It has been suggested that some cases of asthma, chronic fatigue syndrome, fibromyalgia, migraine and depression, may be examples of TILT syndrome. Chemicals such as pesticides may be one of the causes of TILT syndrome, and it is possible that the small amounts in food may serve as the trigger in some people who have lost their tolerance to many chemicals.

Toxicity of Other Components in Pesticide Formulations

While most of the information presented in this chapter relates to the pesticides themselves, it is important to bear in mind that pesticide preparations that are used to spray crops almost always contain other components, some of which may be toxic in their own right. Thus, in addition to the pesticide itself, a particular formulation may contain one or more of a variety of substances, such as solvents, lubricants, solubilisers, suspending agents and preservatives and there have been indications that some of these components may be toxic. In a paper on the toxicity of glyphosate, Japanese researchers have proposed that the wetting agent polyoxyethyleneamine or POEA, a substance used to ensure that glyphosate spray 'wets' the surface of crops (ie completely covers the leaves of crops rather than adhering as globules and then rolling off before it can produce an effect), is the probable cause of many of the toxic symptoms of glyphosate poisoning [10]. The LD50 for POEA is around one third of LD50 for glyphosate, and perhaps this explains the apparent inconsistency, alluded to earlier in this Chapter, between the LD50 for glyphosate and the toxic dose. In addition to the toxic effects of some of the non-pesticide components, it is also possible that there is some sort of interaction between the various components, the scientific term is 'synergism', to produce a mixture that is more toxic than any of the individual components. More recent research has confirmed the toxicity of some of non- glyphosate substances used in formulations on human cells [54].

What Does All This Mean?

The presence of traces of a great variety of pesticides in the food chain has been clearly demonstrated in so many studies as to be now beyond dispute. The issue for us really is whether the often tiny amounts of these substances have had, or are having, any impact on our health and wellbeing. One thing is clear and that is, that despite the apparently rigorous standards laid down by government regulatory bodies before registration and subsequent release of pesticides onto the market, and standards that have been developed to reduce the risk of harm, there is increasing evidence that they do contribute to ill health in at least some sections of the community. This seems to be particularly so in those who are exposed in an occupational setting such as farmers and pest exterminators. This is alarming given that one of the conditions of pesticide release is that they do not harm the users. If the increased risk of a variety of diseases is really related to their exposure to pesticides, as certainly seems likely, it indicates that the results of animal studies may not necessarily be wholly applicable to humans, and that there are other factors which we have not taken into account in setting the permitted exposure rates for humans. Some of these, such as genetic factors, possible greater sensitivity of humans, the synergistic effects of smoking, alcohol, diet and other pollutants, have already been mentioned, but there may be others we know nothing about.

Of course people who do not use pesticides as part of their jobs would appear to have less exposure, at least by inhalation or through the skin, and therefore their risk of disease is lower. However, while exposure may be lower, it cannot be automatically assumed that lower exposure means no risk, because that may depend on the factors mentioned above. Also, while the intake of individual pesticides is low, we know from the various Total Diet Surveys that more than three dozen pesticides have been detected in the average diet, and the total for all of the pesticides combined every day is not so insignificant. And this exposure occurs every day, every month, and every year throughout our lives. What we don't know is whether pesticide exposure through the diet is more or less likely to cause disease in humans than skin and lung contact.

While there is much that is uncertain, there are a number of undisputed facts which could influence many of us to reduce our exposure to pesticides. Firstly, many of the pesticides are poisons with well known effects on critical organs such as the brain. Secondly, in high enough doses they do kill people and in smaller doses they make them very sick and can cause defects in unborn babies. Thirdly, in even smaller doses in the working environment they appear to increase the risk of skin, lung and brain diseases, and cancer. We do not know as yet for certain whether they contribute to degenerative diseases such as cancer in the normal population but there is sufficient suggestive evidence to raise some concerns about their impact on our health. Given that there is a possibility that they are harmful, it certainly makes good sense to reduce our dietary and environmental exposure to these chemicals in whatever way we can.

Summary – Pesticides and Our Health

- There is evidence that acute exposure to pesticides can cause disease and even death.
- Chronic exposure to small amounts of pesticides over time by farmers and others exposed in an occupational setting are believed to increase the risk of certain diseases such as respiratory and skin conditions, Parkinson's disease, reduced mental function and perhaps cancer.
- There is limited evidence for the detrimental effects of pesticide residues in food on human health partly because the required studies are difficult to carry out.
- Susceptibility is a critical factor determining whether a person develops a disease from chronic exposure to the small amounts of pesticides in food.
- Very young children may be particularly susceptible to pesticides because of their small size, and the immaturity of their organs.
- Very young children may be more susceptible to the effects of pesticides because their ability to prevent toxins, such as pesticides, from entering the brain does not develop fully until around 18 months after birth.

- A person's susceptibility to pesticides may be increased if their liver function is impaired through disease or if they are 'slow metabolisers' with a reduced capacity to detoxify and excrete pesticides.
- It is possible that people with chemical intolerance may be susceptible to pesticide residues in food.

References

1. Grube A et al (2011) US Environmental Protection Agency report "Pesticides Industry Sales and Usage: 2006 and 2007 Market Estimates" http://www.epa.gov/opp00001/pestsales/07pestsales/market_estimates2007.pdf
2. US Environmental Protection Agency Report "What are biopesticides" http://www.epa.gov/opp00001/biopesticides/whatarebiopesticides.htm
3. US Environmental Protection Agency "Sources, stressors, and responses- Insecticides" Caddis Volume 2 http://www.epa.gov/caddis/ssr_ins_int.html
4. Huang H et al (2010) Phytotoxicity of sarmentine isolated from long pepper (Piper longum) fruit. J Agric Food Chem 58:9994–10000
5. FDA (2008) Pesticide monitoring program. http://www.fda.gov/Food/FoodSafety/FoodContaminantsAdulteration/Pesticides/ResidueMonitoringReports/ucm228867.htm#FDA_Total_Diet_Study
6. Food Standards Australia and New Zealand (2011) 23rd Total Diet Study. http://www.foodstandards.gov.au/scienceandeducation/publications/23rdaustraliantotald5367.cfm
7. Banday TH et al (2015) Predictors of mortality in organophosphorus poisoning. A case study in rural hospital in Karnataka, India. N Am J Med Sci 7:259–265
8. Lin PC et al (2013) Acute poisoning with neonicotinoid insecticides: a case report and literature review. Basic Clin Pharmacol Toxicol 112:282–286. http://www-ncbi-nlm-nih-gov.proxy.library.adelaide.edu.au/pubmed/23078648
9. Fortenberry GZ et al (2016) Magnitude and characteristics of acute paraquat- and diquat- related illnesses in the US. Environ Res 146:191–199
10. Lee HL et al (2000) Clinical presentations and prognostic factors of glyphosatesurfactant herbicide intoxication: a review of 131 cases. Acad Emerg Med 7:906–910
11. "How Poisonous are Pesticides" (1996) Fact Sheet FS 1/96 Primary Industries South Australia, January 1996
12. Tsai MJ et al (2003) An outbreak of food-borne illness due to methomyl contamination. Toxicol Clin Toxicol 41:969–973
13. Burgess JL et al (2000) Fumigant-related illnesses: Washington State's five year experience. J Toxicol Clin Toxicol 38:7–14
14. Bener A et al (1999) Respiratory symptoms, skin disorders, and serum IgE levels in farm workers. Allerg Immunol 31:52–56
15. Mamane A et al (2015) Occupational exposure to pesticides and respiratory health. Eur Respir Rev 24:306–319
16. Tuchsen F, Jensen AA (2000) Agricultural work and risk of Parkinson's disease in Denmark, 1981-1993. Scand J Work Environ Health 26:359–362
17. Schilter B et al (1996) Limits of pesticide residues in infant foods. A safety based proposal. Regul Toxicol Pharmacol 24:126–140
18. Baltazar MT et al (2014) Pesticides exposure as etiological factors of Parkinson's disease and other neurodegenerative diseases – a mechanistic approach. Toxicol Lett 230:85–103
19. Malek AM et al (2013) Environmental and occupational risk factors for amyotrophic lateral sclerosis: a case control study. Environ Res 117:112–119

20. Sherman JD (1996) Chlorpyrifos (Dursban)-associated birth defects: report of four cases. Arch Environ Health 51:5–8
21. Heeren G et al (2004) Agricultural chemical exposures and birth defects in the eastern Cape Province, South Africa. A case-control study. Environ Health 2:11
22. Mesnage R et al (2010) Two cases of birth defects overlapping Stratton-Parker syndrome after multiple pesticide exposure. Occup Environ Med 67:359
23. Enrique MO et al (2004) Prevalence of fetal exposure to environmental toxins as determined by meconium analysis. Neurotoxicology 23:329–339
24. Meyer A et al (2003) Cancer mortality among agricultural workers from Serrana Region, state of Rio de Janeiro, Brazil. Environ Res 93:264–271
25. Bonner MR et al (2016) Occupational exposure to pesticides and the incidence of lung cancer in the agricultural health study. Environ Health Perspect
26. Ji BT et al (2001) Occupational exposure to pesticides and pancreatic cancer. Am J Ind Med 39:92–99
27. Antwi SO et al (2015) Exposure to environmental chemicals and heavy metals, and risk of pancreatic cancer. Cancer Causes Control 26:158301591
28. Fritschi L et al (2015) Occupational exposure to N-nitrosamines and pesticides and the risk of pancreatic cancer. Occup Environ Med 72:678–683
29. Fleming LE et al (1999) Cancer incidence in a cohort of licensed pesticide applicators in Florida. J Occup Environ Med 41:279–288
30. Koutros S et al (2013) Risk of total and aggressive prostate cancer and pesticide us in the agricultural health study. Am J Epidemiol 177:59–74
31. Flower KB et al (2004) Cancer risk and parental pesticide application in children of agricultural health study. Environ Health Perspect 112:631–635
32. Meinert R et al (2000) Leukemia and non-Hodgkin's lymphoma in childhood exposure to pesticides: results of a register-based case-control study in Germany. Am J Epidemiol 151:639–646
33. Ackermann W et al (2015) The influence of glyphosate on the microbial and production of botulism neurotoxin during ruminal fermentation. Curr Microbiol 70:374–382
34. Jayasumana C et al (2014) Glyphosate, hard water and nephrotoxic metals: are they the culprits behind the epidemic of chronic kidney disease of unknown etiology in Sri Lanka. Int J Environ Res Public Health 11:2125–2147
35. Curl CL et al (2015) Estimating pesticide exposure from dietary intake and organic food choices: the multi-ethnic study of arteriosclerosis. Environ Health Perspect 123:47483
36. Keys A et al (1986) The diet and 15-ye ar death rate in the seven countries study. Am J Epidemiol 124:903–915
37. Juhler RK et al (1999) Human semen quality in relation to dietary pesticide exposure and organic diet. Arch Environ Contam Toxicol 37:415–423
38. Oates L et al (2014) Reduction in urinary organophosphate pesticide metabolites in adults after a week-long organic diet. Environ Res 132:105–111
39. Bradman A et al (2015) Effect of organic diet intervention on pesticide exposures in young children living in low-income urban and agricultural communities. Environ Health Perspect 123:1086–1093
40. Torjusen H et al (2014) Reduced risk of pre-eclampsia with organic vegetable consumption: results from the prospective Norwegian Mother and Child Cohort Study. BMJ 4(9):e: 006143
41. Bradbury KE et al (2014) Organic food consumption and the incidence of cancer in a large prospective study of women in the United Kingdom. Br J Cancer 110:2321–2326
42. Smith-Spangler CI et al (2012) Are organic foods safer or healthier than conventional alternatives? A systematic review. Ann Intern Med 157:348–366
43. Rodier PM (1994) Vulnerable periods and processes during central nervous system development. Env Health Perspect 102(Supplement 2):121–124
44. Schaefer M (1994) Children and toxic substances. Confronting a major public health challenge. Environ Health Perspect 102(Supplement 2):155–156

45. Bruckner TV (2000) Differences in the sensitivity of children and adults to chemical toxicity. The NAS panel report. Regul Toxicol Pharmacol 31:280–285
46. Garrow JS et al (2000) Human Nutrition and Dietetics, 10th ed
47. Antignac JP et al (2016) Country specific chemical signatures of persistent organic pollutants (POP) in breast milk of French, Danish, and Finnish women. Environ Pollut 218:728–738
48. Munoz-Quezada MI et al (2013) Neurodevelopmental effects in children associated with exposure to organophosphate pesticides: a systematic review. Neurotoxicololgy 39:158–168
49. Tanaka E (1999) Update: genetic polymorphism of drug metabolising enzymes in humans. J Clin Pharm Ther 24:323–329
50. Moyer TP et al (2009) Warfarin sensitivity genotyping: a review of the literature and summary of patient experience. Mayo Clin Proc 84:1079–1094
51. Zuniga-Venegas L et al (2015) Determination of the genotype of serum paraoxonase 1 (PON1) status in a group of agricultural and nonagricultural workers in Coquimbo Region, Chile. J Toxicol Environ Health 78:357–368
52. Harari R et al (2010) Neurobehavioural deficits and increased blood pressure in school-age children prenatally exposed to pesticides. Environ Health Perspect 118:890–896
53. González-Alzaga B et al (2014) A systematic review of neurodevelopmental effects of prenatal and postnatal organophosphate pesticide exposure. Toxicol Lett 23:104–121
54. Eriksson P, Talts U (2000) Neonatal exposure to neurotoxic pesticides increases adult susceptibility: a review of current findings. Neurotoxicology 21:37–48
55. Henneberger PK et al (2014) Exacerbation of symptoms in agricultural pesticide applicators with asthma. Int Arch Occup Environ Health 86:423–432
56. Fukuyama T et al (2011) Prior or coinstantaneous oral exposure to environmental immunosuppressive agents aggravates mite allergen-inducced atopic dermatitis-like immunoreaction in NC/Nga mice. Toxicology 289:132–140
57. Lee I et al (2015) Developmental neurotoxic effects of two pesticides: behaviour and neuroprotein studies on endosulfan and cypermethrin. Toxicology 335:1–10
58. Lubick N (2010) Endosulfan's exit: US EPA pesticide review leads to a ban. Science 328:1466
59. CLJ (1974) Manufacture of aldrin is suspended by the EPA. Science 185:601
60. Preisser AM et al (2011) Surprises perilous: toxic health hazards for employees unloading fumigated shipping containers. Total Environ 409:3106–3113
61. Nandipati S, Litvan I (2016) Environmental exposures and Parkinson's disease. Int J Environ Res Public Health 13:8819
62. Su FC et al (2016) Association of environmental toxins with amyotrophic lateral sclerosis. JAMA Neurol 73:803–811
63. http://www.iarc.fr/en/media-centre/iarcnews/pdf/MonographVolume112.pdf
64. https://www.epa.gov/fera/dose-response-assessment-assessing-health-risks-associated-exposure-hazardous-air-pollutants

Chapter 3
The Plastics Revolution

The late twentieth century will no doubt be remembered as the period in which a revolution in information technology really began to transform the world with the advent of the internet, satellite technology, mobile phones and lap top computers, communication with almost any part of the planet was now possible. At around the same time, another revolution, perhaps not as technologically impressive, but nevertheless of great significance, was also taking place. The plastics revolution has resulted in the substitution of simple materials such as paper, cardboard, wood, and aluminium, with plastics. Indeed, such is the pervasiveness of plastic, it is hard to find any area of human activity where plastic is not used. And of course there are good reasons for this. Plastics are easy and cheap to manufacture (if the environmental costs are not included), stable, unbreakable, and light.

The amount of plastic produced globally every year is truly staggering. According to Plastics Europe, the association of plastics manufacturers in Europe, nearly 300 million tonnes were produced globally in 2013, the amount rising almost 50% in a decade [71]. In Europe, almost 40% of all plastic is used for packaging, with large amounts used for building and construction, household and consumer goods, and motor vehicles.

Most people are unaware of the extent to which plastics have really revolutionized our lives. The carpets, floor coverings, furniture, light fittings, kitchen utensils and aids, and mattresses, may contain plastics. Much of our clothing may be made of polyester or nylon which is a type of plastic. It does not end there because when we get into our car, we find plastic in a sizable proportion of the interior (eg floor coverings, front panels), the exterior, and under the bonnet. When we drive to the supermarket, we find plastic in the food packaging, in the shelving, refrigerators, light fittings, and floor coverings. Of course, when we pay for what we have bought we may take our plastic card out of a plastic wallet or plastic handbag, and carry what we have bought to the car in plastic bags, perhaps even walking to the car in shoes with plastic uppers and soles. Back home, we find plastic everywhere both inside the house and in the garden. If we have children or grandchildren, we find it increasingly difficult to find a toy that is not made of plastic. In more recent times, there has been a

A. Poulos, *The Secret Life of Chemicals*,
https://doi.org/10.1007/978-3-030-80338-4_3

huge growth in electronic equipment eg mobile phones, tablets, printers, and computers, almost all of which contain plastics in one form and another.

How Are Plastics Made?

Plastics are made up of materials that are termed "polymers". Polymers are formed by joining together simpler substances termed "monomers". Large numbers of monomers are assembled into polymers, in a fashion similar to that used to make a chain of beads (the polymer) from a number of individual beads (the monomers). There are numerous examples of polymers in living things including our own bodies. Starch is a polymer of the simple carbohydrate glucose, as are cellulose (the covering or frame work of plants) and glycogen (a reserve store of glucose found in our livers). Proteins are also polymers of amino acids but differ from starch, glycogen and cellulose in that there may be 20 or more different amino acids used to make the protein polymer. Similarly, deoxyribonucleic acid (DNA), is a polymer which is found in our genes and is responsible for passing on hereditary characteristics, is made up four different monomers ("nucleotides") joined together to forms chains of thousands or more nucleotides. Many plastics are made by joining together a single monomer rather than many different types. The process of joining together the monomers is termed "polymerisation".

In animals and plants, polymerisation is a highly complex process requiring a number of protein catalysts (a catalyst is a substance bringing about a chemical change without normally undergoing a chemical change itself). The polymerisation of plastic monomers also requires a bit of help, only instead of protein catalysts, other substances, referred to as inititators are used.

What Types of Plastic Are There?

There is a seemingly endless variety of plastics that are used to make an ever expanding list of products. There are two main sorts of plastics – degradable and non-degradable- but within each of these groups there are great number of individual plastics which differ according to the chemical structure of the monomer (in some cases there may be more than one type of monomer) used to make the polymer. During the manufacturing process, other substances (mostly referred to as additives), may be added to alter the properties of the final product. Thus plasticizers are often added to make a softer, more malleable, plastic, while antioxidants are used to reduce the breakdown of the plastic caused by the interaction with oxygen. Another class of additives, the stabilizers, are added to reduce the breakdown caused by the exposure of the plastic to light. The different colours of plastics are due to the addition of chemical colourants.

The various polymers used may be divided into a number of groups depending on their chemical structures. The main groups include the polyolefines (which include the polyethylenes and polypropylenes), the vinyls (polystyrene and polyvinyl chloride are examples), the esters (polyethylene terephthalate and polycarbonate), and the amides (polyamides). Most of these plastics are derived from products of the oil industry.

Most of the above plastics degrade very slowly, mainly because the chemical links that make up the structures of the polymers are very strong requiring harsh conditions to break down, and these are not normally found in the environment. However, there are slight irregularities within the structure of many of these almost indestructible plastics which do provide a molecular weakness which, over time, can lead to the gradual unraveling of the polymers, rather like the unraveling of a chain of beads into the individual beads. Promoters, chemicals that can speed up the breakdown of the so called "non degradable" plastics are often added. Unfortunately these breakdown products are often not themselves further degraded by microorganisms in the soil to very simple products such as carbon dioxide and water. Even if they do, the process may be very slow. To get around this problem of non-biodegradablility (biodegradability means degradation by living organisms such as bacteria, fungi or algae), scientists have invented other plastics, or added other materials at the time of manufacture of the plastic, to render the plastic biodegradable. Some of the main types of non-degradable plastics with an identifying code number, as well as some of their uses (in some cases mixed with other materials), are shown below.

Polyethylene Terephthalate (PET) (code number 1) – beverage containers, clothing, hats, luggage, insulation, carpet, furniture.

High density polyethylene (HDPE) (code number 2) – freezer bags, water pipes, wire and cable insulation, extrusion coatings.

Polyvinylchloride (V) (code number 3) – building materials, magnetic cards, gramophone records, clothing, upholstery, flexible hoses, flooring, electric cable insulation.

Low density polyethylene (LDPE) (code number 4)- supermarket bags, plastic bottles, cling wrap, car covers, tank and pond liners, moisture barriers.

Polypropylene (PP) (code number 5) – food packaging, textiles, laboratory equipment, automobile components, tubing.

Polystyrene (PS) (code number 6- electrical appliances, electronics components, building materials, laboratory products, food packaging, children's toys.

All other non-degradable plastics have the code number 7. These include numerous others including poly(methylacrylate) (PMMA- also known as perspex), Teflon, polyketone, phenolics (which include such well known materials as Formica and Bakelite).

What Are Degradable Plastics?

The alarming build-up of plastics in the environment has provided much of the impetus for the development of plastics that degrade much faster than the conventional plastics. Essentially, the degradable plastics are divided into two main groups. One of these groups which are referred to simply as "degradable" plastics, contain added substances (also called promoters) that hasten the breakdown of the polymers, and they achieve this by working away at the structural weaknesses mentioned above. The metals iron, nickel, cobalt and antimony, in the presence of ultraviolet light, are examples of substances that can promote the degradation of plastics. Another class of substances, containing what the chemists call a "carbonyl group" (really a combination of atoms of carbon and oxygen), can also do the same thing in the presence of light.

A second group of degradable plastics includes the "biodegradable" plastics. There are a number of these, indeed the number is growing and will continue to grow in the future. They include plastics based on starch, a natural polymer made from the sugar glucose, found in some plants such as corn and potatoes [53]. As starch is soluble in water, plastics made from starch tend to absorb water and swell which limits their use. To counter this, scientists have either chemically modified starch or mixed it with other polymers. Another way is to use starch to make other substances, for example lactic acid, and then use lactic acid as the monomer to make the polymer referred to as "polylactate". Lactic acid is a naturally occurring substance, it is formed in our own bodies from the major fuel driving our body processes, the carbohydrate glucose, and can be converted into a polymer. Bacteria can also be used to make other polymers, such as PHA (or to use the chemical term "polyhydroxyalkanoate") and PHB (polyhydroxybutyrate). This group of plastics is termed the "polyesters" (excluding starch which is not a polyester) and includes many others that are not made by bacteria (PGA and PHBV) to name just a couple.

Microorganisms can act on the biodegradable plastic without the need for promoters in the plastic to start the degradative process because the chemical bonds found in the polymers are not all that different from similar chemical bonds found in the microorganisms' natural environment. In contrast, conventional plastics contain chemical structures that are mostly alien to these same microorganisms and therefore, from an evolutionary point of view, until very recently, there has never been the need to develop a way of utilizing chemicals that are not a part of their environment. However, microorganisms are extremely adaptable and it was believed that some may have developed mechanisms to handle these types of chemical bonds. Scientists looking for such organisms have found that there are a number of microorganism species that can degrade one of the most chemically stable of the apparently non-degradable plastics, polyethylene [1, 2, 54]. Indeed, one doomsday scenario envisages the spread of bacterial strains that can utilize certain types of plastic as a nutrient. Because of the fact that plastic is everywhere, it is easy to see how devastating the spread of such bacterial strains would be. While there are bacteria that can utilize the components of the so-called non-degradable plastics, these are the exceptions rather than the rule and, as a consequence, this type of plastic may remain in landfills for many years with little change.

What Does Plastic Contain?

The various polymers are mostly unsuitable, by themselves, for the many, and diverse, uses that are required for plastics. They are either too hard, or soft, or brittle. The solution developed many years ago was to add other chemical substances, termed additives, to change the properties of the polymer. Thus most plastics contain the basic polymer or, in some cases mixtures of polymers, together with other chemicals. Because of the concerns of contamination, plastic used for food packaging has a rather limited range of additives. On the other hand, the number and variety of additives included in plastics used for other purposes is truly astounding. To illustrate the sorts of substances present in non-food plastic, it is worthwhile looking at what has been reported to be present in plastic toys [3]. In a study published in July 2005, 113 samples of plastic toys were analysed. The main plastics used were PVC, acronitrile butadiene styrene copolymer (a copolymer is a mixture of polymers), polyethylene, polypropylene, and ethylene vinyl acetate. A list of some of the main types of additives found in these samples is shown below.

Monomers	Light stabilisers
Oligomers	Fluorescent whitening agents
Inititators	Plasticisers
Antioxidants	Lubricants
Dyes and inks	Antimicrobials
Flame retardants	Heat stabilisers

Why Are There Additives in Plastic?

The presence of small amounts of monomers (the starting materials used to make the polymer) in many of the plastics is not surprising, nor is it surprising that there are fragments of the polymers (oligomers) present. Antioxidants, whitening agents, and light stabilizers are added to reduce the effects of oxygen in the air, light and heat, on the plastic both during and after manufacture. The effects relate mainly to loss of strength, stiffness, and discolouration. Dyes and inks are used to colour the finished plastic, while plasticizers serve to soften the plastic, thereby increasing its flexibility. Flame retardants are added to plastics, particularly for those used in building, motor vehicles, and electronic appliances, to reduce the risk of fires and generation of smoke. Lubricants reduce the effects of friction as well as enhancing the release of plastics from molds during the manufacture of plastic goods. Antimicrobials are sometimes added to prevent the odour, staining and discolouration that can arise from the growth of bacteria, algae and fungi on the surfaces of plastic items.

It is important to note that the above list is not complete because there are other types of additives, and some of these have quite specific functions when used with a particular plastic. It is also worth noting that within each additive type there may be a great number of separate and distinct chemical substances. A couple of examples illustrate this point very well. Plasticizers, which may make up as much as 20% of the final plastic, include the phthalate-based plasticizers (there are at least 10 or more to choose from), the adipate-based plasticizers (at least four of these), trimellitates (five or more), maleates, sebacates, and benzoates. Flame retardants include the brominated (there may be dozens to choose from), chlorine, and phosphorus based retardants. The choice of additives available to manufacturers is truly staggering. A good example of the nature, and complexity, of the additives present in plastics has been well illustrated by the analysis of a variety of plastic marine debris which has shown the presence of 100 or more different chemicals [48].

Nanomaterials

More recently, yet another class of additives, the so-called "nanomaterials" have begun to be introduced into plastics. Nanomaterials are those substances or materials that have been reduced in size down to the nano level, a size much closer to the dimensions of the atoms and molecules that make up all of matter. For example, the terms "nanogram" refers to a mass that is one thousandth of a million gram, while a "nanometer" is a thousandth of a millionth of a metre. Nanotechnology refers to a technology utilizing the special properties of ultra small particles. In relation to one of its many uses, nanomaterials are being introduced into plastics [49]. Examples include nano-clay, added to plastic packaging as a barrier to reduce the movement of air into, and loss of water out of, a packaged food, and nano-silver, which acts as an anti-bacterial agent. There seems little doubt that small amounts of these newer plastic components migrate into food and also released into the environment in landfills.

What Happens When Plastic Degrades?

In the case of the non-degradable plastics, the combined effects of light (particularly UV light), heat, water, and soil components, chips away and, over time, cracks may develop in the plastic and small pieces fall off. The smaller pieces are perhaps more vulnerable to degradation because of the bigger surface area exposed to the elements which perhaps explains the large amounts of what have been termed "microplastics', that is plastic degradation products of varying particle size, many less than 0.1 millimetre ie 1/100 th of a centrimetre found in the marine environment

[72]. Microplastics can form more quickly through the washing of clothes containing plastic, polyester fabrics are an example. The combined effects of water, detergent and agitation in washing machines releases significant amounts of microplastics [73]. These degradative processes also result in the release of the various additives than have been added to the plastic during the manufacture. While the polymers from non-degradable plastics (eg polyethylene, PET, etc) are resistant to attack by microbes in the soil and water (although as pointed out above there are some species of microorganism that can slowly degrade them), the various additives can be utilized by microorganisms. On the other hand, the disintegration of degradable plastic which contains promoters will proceed much more rapidly particularly if exposed to light. The plastic will gradually fall apart releasing any additives and polymers. The breakdown of biodegradable plastic also proceeds quite rapidly but in this case degradation is brought about by microbial action on the polymer itself.

Plastic degradation can occur in landfills, although it is believed that this process is slower because of the absence of light and heat. As mentioned above, there are microbes that degrade many of the plastics and, while the breakdown of the polymers is slow, the additives, when released from the plastic are degraded faster. Plastics can also be burnt. The simpler plastics, such as those that contain polyethylene and polypropylene, produce mainly carbon dioxide and water. However, if there are any additives in the plastic, or if anything else is burnt with them, other substances will also be produced [55]. While the identity of some of these is known, what is formed is not always predictable and nor do we know all of the likely products. In the case of the chlorine containing plastics eg PVC, some particularly nasty emission products can be produced, depending on the temperature of burning. The most infamous of these are the cancer producing dioxins. Burning therefore releases many different chemical substances which are carried by air currents and eventually deposited either on land, the sea or in rivers.

How Can the Chemicals in Plastics Be Taken Up into Our Bodies?

There are numerous reports that show that plastic components, particularly the additives, are present in our blood and urine indicating that they can be taken up into our bodies [67]. There are a number of potential ways they can do this—via the skin, mouth, nose, and even through our blood. Skin contact is very easy as it involves simple handling, something we do without thinking many times in a day with the many different plastic items that are now such an integral part of our life. Of course, if some of the plastic components do rub off onto our skin, particularly our hands, they can also find their way into our mouth and from there into our stomachs. Skin is not a complete barrier to the entry of foreign substances in our bodies. Many substances can cross this barrier and enter the blood. Some fatty substances can certainly move from the skin into the blood, and there indications that some of the additives, many of which, such as the fire retardants have some of the characteristics of fats, can enter the body through the skin, so it is likely that some of the additives in particular can cross through the skin layers and eventually find their way into our blood.

Some of the additives, most notably the plasticizers and the flame retardants are relatively volatile and can be detected even in the air we breathe, particularly indoors where they are probably released from all sorts of items including floor coverings, furniture, floor polish etc. and so can enter our bodies in this way [4, 5, 50]. Another source of air exposure possibly occurs in motor vehicles. With the sun beating down, temperatures can rise to in excess of even 50 degrees Centigrade and, under these conditions, fire retardants and plasticizers can vaporize and be breathed into the lungs [39]. The release of something into the air is obvious to anyone getting into a car that has been left in the hot sun for a long time. Volatile plastics components can also be released into the air that passes through the plastic tubing in respirators that are used for treating people with respiratory diseases [6].

The plastic catheters and tubing used for blood transfusions and kidney dialysis are also a source of exposure to the various additives present in the plastic, often PVC, and can enter the blood directly thence to our various organs [7–9].

Another way we are exposed to the additives in plastics is oral, either through eating food that contains additives or, in the case of babies for example, via the sucking and chewing of toys and other plastic items [40, 41]. Food is an important source of additives, and mostly as a consequence of the use of plastic packaging for storage. The release of plasticizers in particular packaging materials is a well known phenomenon [77, 79]. Plasticizers are readily released particularly if the plastic is used for foods that contain high amounts of fats, such as vegetable oils [42]. Small amounts of monomer, left over from the synthesis of the polymer, and substances referred to as oligomers (a number of monomers joined together), initiators, antioxidants and anything else that was added during manufacture, or contaminants in the raw material that was used to manufacture the plastic may have the potential to migrate into the food. And migration of traces of various substances may occur even into non fat foods such as drinks and even water [10, 11, 51].

We can illustrate the point by using one of the most apparently chemically inert plastics. PET is being used increasingly to store food and drink. Analysis of PET itself has confirmed that there may be up to 19 different chemical substances in the plastic with the potential to migrate into food or water [10]. At least some of these substances have been shown to migrate into water or into food simulants [11, 12]. The latter are used instead of food to determine the likely migration of the additives because it is much easier chemically to detect contaminants in the simulant than in food which is much more chemically complex.

Even babies' food – breast milk – is contaminated with a variety of plasticizers, presumably entering the mothers' bodies via the sources mentioned above, and then gradually finding their way into their milk during lactation. Danish, Italian, and Korean researchers have all detected easily measurable amounts of a variety of phthalates in human breast milk [13, 28, 52].

Flame retardants, which have been mentioned in another chapter, are also found in various plastics [14, 15]. There are numerous reports confirming the presence of one of the major group of flame retardants, PBDEs, in food and this has been discussed elsewhere in this book [16]. There are indications that food is a relatively minor source and that household and other dusts, including electronic equipment such as televisions, furniture, and even lint from dryers, may represent the major source of the PBDEs found in the human body [17–19]. It has been speculated that PBDEs can enter the body by either inhalation, ingestion or skin contact.

Are Plastic Components Harmful?

There is now quite a lot of information available demonstrating the presence of a variety of plastic components in human blood, urine and tissues. Phthalates and bisphenol A have caused the greatest concern because these substances are known to have endocrine disrupting properties (EDCs or endocrine disrupting compounds). That is, they can interfere with the actions of many of the body's hormones, especially the sex hormones [32].

Reproduction and Growth

Animal studies have pointed to the ability of the plasticizers, and particularly the phthalates and bisphenol A, to interfere with growth and development. Thus exposure of pregnant rats to some phthalates, has been observed to produce a variety of effects on the reproductive tract in the male offspring including hypospadias (abnormality of the penis), and cryptorchidism (an abnormality of the testis, the organ which produces sperm) [20, 21, 56]. While the most reported effects relate to abnormalities in the male sexual organs of rats, there are also indications that prolonged exposure can lead to the development of cancers in the liver and testis [22, 56].

There are suggestions that the phthalates may also increase the likelihood of developmental abnormalities in humans. For example, a group of US researchers have claimed that the anogenital distance ie the distance between the anus (ie the opening through which the faeces pass) and the base of the penis in boys is reduced if their mothers were exposed before birth to certain phthalates [57, 69, 70]. It has been argued that changes in the anogenital distance, which have also been shown in animals exposed to phthalates, are an indication of an abnormality in sexual development. Phthalates may also increase the risk of cryptorchidism in humans [30]. There are suggestions as well that phthalates may affect sexual development in girls by inducing earlier breast development, and in boys by delaying puberty perhaps by interfering with the actions of sex hormones [23, 59, 60, 62]. However these findings remain controversial because other researchers have found no such effects [44]. There are also indications that exposure to phthalates may affect the children of those exposed. Thus, exposure to dibutylphthalate, through repeated application to clothing to prevent tick infections, is thought to have caused an increased risk of abnormalities in sexual development of children born to the fathers of those exposed [36]. This possible transgenerational effect of dibutylphthalate is reminiscent of the effect of diethylstilbestrol (DES) on the daughters and grandsons of women exposed during pregnancy [45]. However, once again, these findings have been disputed [61].

A recent review of much of the research carried out over the last decade on the effects of phthalates on animal and human health has confirmed the effects on animals, particularly rats. Much of the research on animals generally involves the use of much higher concentrations than those to which humans are exposed. Despite this, there are genuine grounds for concern for a number of reasons. Firstly, phthalates most definitely do have endocrine disrupting properties so they are potentially harmful. Secondly, there is some research which supports their ability to interfere with growth and development. Finally, much of our exposure is to complex mixtures of phthalates, together with other environmental contaminants rather than as single substances and we really know little about whether they are more or less active in this form [62].

Phthalates are not the only plastic components that may have effects on growth and reproduction. Bisphenol A, like phthalates, also has endocrine disrupting properties because it binds to some of the same receptors that are used by estrogens, the female sex hormones. Receptors are essential for hormone activity because the binding of hormones to its individual receptors is an essential first step in a cascade of reactions culminating in effects on DNA and, finally, the production of specific proteins. It is easy to see how binding of bisphenol A to an estrogen receptor can interfere with the action of these hormones.

Animal experiments have suggested a possible link between exposure to bisphenol A and effects on sexual development and growth eg the development of the mammary gland (the milk producing gland), earlier age of onset of puberty, and birth weight [32]. There is evidence that the concentrations of bisphenol A detected in human blood and tissues, as well as the blood of fetuses, is sufficient to cause behavioral abnormalities in mice [17, 18]. The sperm counts and sexual function in men occupationally exposed to bisphenol A have been reported to be depressed but many other effects in humans have been disputed [33, 35, 67]. This particular report was highly critical of another, funded by American Plastics Council, which concluded that low doses produced little measurable effects. Some evidence that exposure, as determined by the measurement of the levels of bisphenol A in the urine, is linked to degenerative diseases such as diabetes and cardiovascular diseases, as well as abnormalities in liver function and obesity and abnormalities of the ovaries in women [19, 20]. On the other hand, the US government agency, the FDA, believes that "the scientific evidence at this time does not suggest that the very low levels of human exposure to BPA through the diet are unsafe" [26]. The FDA report has in turn been criticized for its preference for industry-funded rather than government funded research, the argument being that industry-funded research may be biased [12]. There is Despite this, the US FDA early in 2010 stated that it had "some concern about the potential effects of BPA (bisphenol A) on the brain, behavior, and prostate gland in fetuses, infants, and young children" even though later it claimed that it was safe [77]. It is worthwhile noting as well that, despite the FDA's insistence that bisphenol A is safe, some manufacturers have discontinued its use in infant formula packaging.

some evidence that exposure, as determined by the measurement of the levels of bisphenol A in the urine, is linked to degenerative diseases such as diabetes and cardiovascular diseases, as well as abnormalities in liver function and obesity and abnormalities of the ovaries in women [19, 20]. On the other hand, the US government agency, the FDA, believes that "the scientific evidence at this time does not suggest that the very low levels of human exposure to BPA through the diet are unsafe" [26]. The FDA report has in turn been criticized for its preference for industry-funded rather than government funded research, the argument being that industry-funded research may be biased [12, 78].

Diabetes, Obesity, and Heart Disease

Phthalates may also interfere with the body's ability to handle the principal fuel of the body ie the sugar glucose. The hormone insulin is required for the utilization of glucose by the various organs of the body. If it is missing, or present in smaller than normal amounts such as in the childhood form of diabetes, also called juvenile or type 1 diabetes, glucose accumulates in the blood leading to "hyperglycaemia" (meaning excess glucose in the blood). Different groups of researchers have shown an apparent relationship between the levels of phthalates in urine (an indication of exposure to phthalates), what is termed "insulin resistance" and the" metabolic syndrome" in humans [25, 38, 63]. Insulin resistance refers to an inability to respond appropriately to insulin and this results in a rise in blood sugar or hyperglycemia, a feature of diabetes. Individuals with the metabolic syndrome have increased blood glucose and cholesterol, blood pressure and body fat. There are some reports linking phthalates to the development of obesity but there is limited evidence that these substances increase the risk of heart disease [29, 64].

Bisphenol A may also have a similar, although broader effect. For example, in a study published in 2008 it was shown that mice injected with relatively small amounts of bisphenol A (ten millionths of a gram per kg of body weight) produced an outpouring of insulin from the pancreas (the organ that produces insulin which regulates blood sugar). At the same time, there was a reduction in the levels of blood glucose, the main fuel of the body [24]. At higher doses, bisphenol increased the amount of insulin produced by pancreas and also made the animals insulin resistant. In humans, the combination of hyperglycemia and normal levels of insulin is a hallmark of type 2 diabetes and the reasons for this are unknown. Although the experiments described were carried out on mice, and therefore may not be relevant to humans, there is some evidence that bisphenol A exposure may increase the risk of obesity and insulin resistance in humans [65, 66]. There are suggestions as well of a relationship between bisphenol A and the risk of heart disease [34, 46].

Because of possible adverse effects of bisphenol A, alternatives have been sought and are being used increasingly. Two of these, related chemically to bisphenol A, are bisphenols S and F. However, both of these show similar endocrine disrupting properties [47].

While much of the focus up to now has been on plasticizers and monomers, there have been limited reports on whether any or all of the many other components find their way into our bodies and, if they do, whether they cause harm. Certainly we do know that some of these, for example the UV stabilizers and antioxidants can migrate into food, food simulants, and drinks so it would be surprising indeed if they do not get into our bodies [12, 26] That they may have the potential to harm is indicated by reports that UV stabilizers such as some of the benzophenones, which are used in sunscreens and plastics, are estrogenic ie they can mimic the effects of the female sex hormones [68].

What Does All This Mean?

Our exposure to a variety of plastics and their components, in particular the plasticizers, and the ability of these substances to enter our bodies in different ways is beyond dispute. The real issue is not whether some of the plasticisers are potentially harmful because certainly animal studies have shown consistently that in larger amounts they are, and there is no real reason to believe that human tissues would respond all that differently from rodent tissue, the most tested animal species. After all, when the safety of new chemicals are assessed, they are often first tested in rodents and later in other animals such as dogs and perhaps monkeys, Rodents are also used during the early stages of drug development. If a potential drug can be shown to work in rodents then it increases the likelihood that it will be efficacious in humans. While there are clear differences between humans and rodents, there are also similarities in the structure and chemical composition of the major organs such as the liver, brain and kidneys. Of course there are differences as well and that is why other animal species are also used. What we do not know is whether the small amounts of the various chemical derived from plastics, amounts much smaller than those shown to be harmful in rodents, represent a hazard. There are hints that even in these small amounts they may cause harm in humans, mainly by interfering with the actions of natural hormones. Unfortunately the jury is still out on this and, meanwhile, we continue to be exposed to these substances.

References

1. Hadad D et al (2005) Biodegradation of polyethylene by the thermophilic bacterium Brevibacillus borstelensis. J Appl Microbiol 98:1093–1100

2. Yang J et al (2014) Evidence of polyethylene biodegradation by bacterial strains from the guts of plastic-eating waxworms. Environ Sci Technol 48:13776–13784

3. Netherlands Food and Goods Authority Report (2005) "Screening of plastic toys for chemical composition and hazards"

4. Marklund A et al (2005) Organophosphorus flame retardants and plasticizers in air from various indoor environments. J Environ Monit 7:814–881

5. Erythropel HC et al (2014) Leaching of the plasticizer di(2-ethylhexylphthalate) (DEHP) from plastic containers and the question of human exposure. Appl Microbial Biotechnol 98:9967–9981

6. Hill SS et al (2003) Plasticisers, antioxidants, and other contaminants delivered by PVC tubing used in respiratory therapy. Biomed Chromotagr 17:250–262

7. Haishima Y et al (2004) Risk assessment of di-2-(ethylhyxyl)phthalate released from PVC blood circuits during hemodialysis and pump-oxygenation therapy. Int J Pharm 274:119–129

8. Green S et al (2005) Use of di(2-ethylhexyl) phthalate-containing medical products and urinary levels of mono(2-ethylhexyl) phthalate in neonatal intensive care unit infants. Environ Health Perspect 113:1222–1225

9. Tickner JA et al (2001) Health risks posed by use of di-2-ethylhexyl phthalate (DEHP) in PVC medical devices: a critical review. Am J Ind Med 39:100–111

10. Kim H et al (1990) Determination of potential migrants from commercial amber polyethylene terephthalate bottle wall. Pharmacol Res 7:176–179

11. Mutsuga M et al (2006) Migration of formaldehyde into mineral water in polyethylene terephthalate (PET) bottles. Food Addit Contam 23:212–218

12. Begley TH et al (2004) Migration of a UV stabilizer from polyethylene terephthalate into food simulants. Food Addit Contam 21:1007–1014

13. Main KM et al (2006) Human breast milk contamination with phthalates and alterations of endogenous reproductive hormones in infants three months of age. Environ Health Perspect 114:270–276

14. Anh HQ et al (2016) Polybrominated diphenyl ethers in plastic products, indoor dust, sediment and fish from informal e-waste recycling sites in Vietnam: a comprehensive assessment of contamination, accumulation pattern, emissions, and human exposure. Environ Geochem Health. 2016 Aug 19. [Epub ahead of print]

15. Vojta S et al (2017) Screening for halogenated flame retardants in European consumer products, building materials, and wastes. Chemosphere 168:457–466

16. Schecter A et al (2008) Brominated flame retardants in US food. Mol Nutr Food Res 52:266–272

17. Schecter A et al (2009) PBDEs in US and German clothes dryer lint: a potential source of indoor contamination. Chemosphere 75:623–628

18. Lorber M (2008) Exposure of Americans to polybrominated diphenyl ethers. J Expo Sci Environ Epidemiol 18:2–19

19. Allen JG et al (2008) Linking PBDEs in house dust to consumer products using X-ray fluorescence. Environ Sci Technol 42:4222–4228

20. Foster PM (2006) Disruption of reproductive development in male rat offspring following in utero exposure to phthalate esters. Int J Androl 29:140–147

21. Andrade AJ et al (2006) A dose response study following in utero and lactational exposure to di-(2-ethylhexyl) phthalate (DEHP): reproductive effects on adult male offspring rats. Toxicology 228:85–97

22. Voss C et al (2007) Lifelong exposure to di-(2-ethylhexyl) –phthalate induces tumours in liver and testes of Sprague-Dawley rats. Toxicology 17:259–277

23. Colon I et al (2000) Identification of phthalate esters in the serum of young Puerto Rican girls with premature breast development. Environ Health Perspect 108:895–900

24. Ropero et al (2008) Bisphenol A disruption of the endocrine pancreas and blood glucose homeostasis. Int J Androl 31:194–200

25. Stahlhut RW et al (2007) Concentrations of urinary phthalate metabolites are associated with increased waist circumference and insulin resistance in adult U.S. males. Environ Health Perspect 115:876–882
26. Simoneau C et al (2012) Identification and quantification of the migration of chemicals from plastic bottles used as substitutes for polycarbonate. Food Addit Contam Part A Chem Anal Control Risk Assess 29:469–480
27. Schecter A et al (2003) Polybrominated diphenyl ethers (PBDEs) in US mother's milk. Environ Health Perspect 111:1723–1729
28. Latini G et al (2009) Lactational exposure to phthalates in Southern Italy. Environ Int 35:236–239
29. Oktar S et al (2015) The relationship between phthalates and obesity: serum and urinary concentrations of phthalates. Minerva Endocrinol
30. Main KM et al (2010) Genital abnormalities in boys and the environment. Best Pract Res Clin Endocrinol Metab 24:279–289
31. Li DK et al (2010) Relationship between urine bisphenol-A level and declining male sexual function. J Androl 31:500–506
32. Ranjit N et al (2010) Bisphenol-A and disparities in birth outcomes: a review and directions for future research. Perinatol 30:2–9
33. Lang IA et al (2008) Association of urinary bisphenol A concentration with medical disorders and laboratory abnormalities in adults. JAMA 300:1303–1310
34. Melzer D et al (2010) Association of urinary bisphenol a concentration with heart disease: evidence from NHANES 2003/2006. PLoS One 13:e8673
35. Li DK et al (2010) Relationship between urinary bisphenol-A level and declining sexual function. J Androl 31:500–506
36. Carran M, Shaw (2012) New Zealand Malayan war veterans' exposure to dibutylphthalate is associated with an increased incidence of cryptorchidism, hypospadias, and breast cancer in their children. NZ Med J 125:52–63
37. Brouwers MM et al (2007) Risk factors for hypospadias. Eur J Pediatr 166:671–678. http://www.ncbi.nlm.nih.gov/pubmed/17103190
38. Huang T et al (2014) Gender and racial/ethnic differences in the associations of urinary phthalate metabolites with markers of diabetes risk: National Health and Nutrition Examination Survey 2001–2008. Environ Health 13:6
39. Geiss O et al (2012) Investigation of volatile organic compounds and phthalates present in the cabin air of used private cars. Environ Int 35:1188–1195
40. Johnson S et al (2011) Phthalates in toys available in the Indian market. Bull Environ Contamin Toxicol 86:621–626
41. Correa-Tellez KS et al (2008) Estimated risk of water and saliva contamination by phthalate diffusion from plasticized polyvinyl chloride. Environ Health 71:34–39
42. Xu Q et al (2010) Analysis of phthalate migration from plastic containers to packaged cooking oil and mineral water. J Agric Food Chem 58:11311–11317
43. Kay VR et al (2014) Reproductive and developmental effects on phthalate diesters in males. Crit Rev Toxicol 44:467–498
44. Wolff MS et al (2014) Phthalate exposure and pubertal development in a longitudinal study of US girls. Hum Reprod 29:1558–1566
45. Kalfa N et al (2011) Prevalence of hypospadias in grandsons of women exposed to diethylstilbestrol during pregnancy: a multigenerational national cohort study. Fertil Steril 39:2574–2577
46. Ranciere F et al (2015) Bisphenol A and the risk of cardiometabolic disorders: a systematic review with meta-analysis of the epidemiological evidence. Environ Health 14:46
47. Rochester JR, Bolden AL (2015) Bisphenol S and F: a systematic review and comparison of the hormonal activity of bisphenol A substitutes. Environ Health Perspect 123:643–650
48. Rani M et al (2015) Qualitative analysis of additives in plastic marine debris and its new products. Arch Environ Contam Toxicol 69:352–366

49. Nanotechnologies in Food Packaging: an Exploratory Appraisal of Safety and Regulation. Food Standards Australia and New Zealand, June 2016. http://www.foodstandards.gov.au/publications/Pages/Nanotechnologies-in-Food-Packaging-an-Exploratory-Appraisal-of-Safety-and-Regulation.aspx
50. Xu Y et al (2015) Semi-volatile organic compounds in heating, ventilation, and air-conditioning filter dust in retail stores. Indoor Air 25:79–92
51. Guo Z et al (2010) Determination of six phthalic acid esters in orange juice packaged by PVC bottle using SPE and HPLC-UV: application to the migration study. J Chromatogr Sci 48:760–765
52. Kim S et al (2015) Concentrations of phalate metabolites in breast milk in Korea: estimating exposure to phthalates and potential risks among breast-fed infants. Sci Total Environ 508:13–19
53. Ferreira AR et al (2016) Polysaccharide-based membranes in food packaging applications. Membranes 6:22
54. Krueger MC et al (2015) Prospects for microbiological solutions to environmental pollution with plastics. Appl Microbiol Biotechno 99:8857–8874
55. Tue NM et al (2016) Release of chlorinated, brominated and mixed dioxin-like related compounds to soils from open burning of e-waste in Agbogbloshie (Accra, Ghana). J Hazard Mater 302:151–157
56. Zarean M et al (2016) A systematic review on the adverse health effects of di-2-ethylhexyl phthalate. Environ Sci Pollut Res Int 23:24642–24693
57. Swan SH et al (2015) First trimester phthalate exposure and anogenital distance in newborns. Hum Reprod 30:963–972
58. Jensen TK et al (2016) Prenatal exposure to phthalates and anogenital distance in male infants from a low-exposed Danish cohort. Environ Health Perspect 124:1107–1113
59. Zhang Y et al (2015) Could exposure to phthalates speed up or delay pubertal onset and development? A 1-5 year follow up of a school-based population. Environ Int 83:41
60. Frederiksen H et al (2012) High urinary phthalate concentrations associated with delayed pubarche in girls. Int J Androl 35:216–226
61. McBride D, Schep L (2012) Comment on Carran and Shaw's "New Zealand Malayan war veterans' exposure to dibutylphthalate" article. NZ Med J 125:105–106
62. Mariana M et al (2016) The effects of phthalates in the cardiovascular and reproductive systems: a review. Environ Int 94:758–776
63. James-Todd TM et al (2016) The association between phthalates and metabolic syndrome: the National Health and Nutrition Examination Survey 2001–2010. Environ Health 15:52
64. Stojanoska MM et al (2016) The influence of phthalates and bisphenol A on the obesity development and glucose metabolism disorders. Endocrine
65. Hong SH et al (2016) Urinary bisphenol A is associated with insulin resistance and obesity in reproductive-age women. Clin Endocrinol
66. Provvisiero DP et al (2016) Influence of bisphenol A on type 2 diabetes. Int J Environ Res Public Health 6
67. Giulivo M et al (2016) Human exposure to endocrine disrupting compounds. Their role in reproductive systems, metabolic syndrome and breast cancer. A review. Environ Res 151:251–264
68. Kang HS et al (2016) Urinary benzophenone concentration and their association with demographic factors in a South Korean population. Environ Res 149:1–7
69. Swan SH et al (2005) Decrease in anogenital distance among male infants with prenatal phthalate exposure. Environ Health Perspect 113:1056–1106
70. Martino-Andrade AJ et al (2016) Timing of prenatal phthalate exposure in relation to genital endpoints in male newborns. Andrology 4:585–593
71. Plastics Europe (2014/2015) Plastics – the facts. An analysis of European plastics production, demand and waste data http://www.plasticseurope.org/documents/document/20150227150049-final_plastics_the_facts_2014_2015_260215.pdf

72. Bouwmeester H et al (2015) Potential health impact of environmentally released micro- and nanoplastics in the human food production chain. Experiences from nanotechnology. Environ Sci Technol 49:8932–8947
73. Hernandez E et al (2017) Polyester textiles as a source of microplastics from households: a mechanistic study to understand microfiber release during washing. Environ Sci Technol 51:7036–7046
74. vom Saal FS, Hughes C (2005) An extensive new literature concerning low dose effects of bisphenol A shows the need for a new risk assessment. Environ Health Perspec 113:926–933
75. Palanza P et al (2008) Effects of development exposure to bisphenol a on brain and behavior in mice. Environ Res 108:150–157
76. Takeuchi T et al (2004) Positive relationship between androgen and the endocrine disruptor, bisphenol A, in normal women and women with ovarian dysfunction. Endocr J 51:165–169
77. USFDA Draft assessment of bisphenol A : Use in food and contact applications. 2008a
78. Myers JP et al (2009) Why public health agencies cannot depend on good laboratory practice for selecting data: the case of bisphenol A. Environ Health Perspect 117:309–315. http://www.ncbi.nlm.nih.gov/pubmed/19337501
79. Biedermann M, Grob K (2013) Assurance of safety of recycled paperboard for food packaging through comprehensive analysis of potential migrants is unreasistic. J Chromatog A 1293:107–119

Chapter 4
Toxic Metals

Metals such as iron, copper, and tin have been a necessary part of human existence for millennia. Even arsenic, lead, and mercury, three very poisonous metals, have also had widespread use through the ages. Though their toxicity was well known, even to the ancients, it did not prevent their use even as therapeutic agents and ingredients of cosmetics. In more recent times, however, the use of these metals, and their applications particularly in industry has extended into areas undreamed of by the ancients. Industrialisation has almost certainly played a major role in the variety of their uses while, more recently, globalization has ensured that people everywhere enjoy the benefits of their often very special properties. However, as is seen repeatedly for the burgeoning number of chemicals in our life, their special properties, ie what makes them so attractive for industry, have to be balanced against their toxicity and therefore their capacity to harm. It has been assumed – and this is the argument that is used for all of the environmental pollutants that we are exposed to – that the amounts released into the environment are too small to constitute a health hazard. In this chapter we will examine this proposition in some detail.

Coal Combustion as a Major Source of Environmental Heavy Metal Contamination

Most people would be surprised to learn that coal contains significant amounts of lead, mercury and arsenic [54–56]. The levels of the various metals seem to vary according to the source of the coal. Combustion of coal in power stations releases some of the metals into the air, and the fly ash, the material remaining after combustion, also contains appreciable amounts. Depending on the method used, disposal of fly ash can lead to the release of these metals into the environment. Governments have put pressure on the energy industry to introduce new technologies, particularly in the developed world, to reduce the amounts released into the

© The Author(s), under exclusive license to Springer Nature Switzerland AG 2021
A. Poulos, *The Secret Life of Chemicals*,
https://doi.org/10.1007/978-3-030-80338-4_4

environment but the combustion of coal remains an important source of environmental contamination with heavy metals [57, 58]. However, coal is not the only source of heavy metal contamination – there are many - and these will be discussed in this chapter [45].

Lead

The variety of uses of lead in industry is very great. From the well-known lead batteries, to lead sheets and pipes (lead pipes these days are mainly used for corrosive liquids), cable sheaths, stained glass windows, solders, radiation shielding, plastics, glass, ceramics, road markings to name just a few. The use of lead in paint has been phased out in most developed countries except perhaps for rust inhibiting paints, although it is still used in many parts of the world [1]. It has also been phased out as a petrol additive in developed countries and in an increasing number of developing countries.

Sources of Exposure

Until recently, probably the greatest exposure to lead occurred as a consequence of its addition, in the form of lead tetraethyl and other related lead containing substances, to petrol as an anti-knock additive. Put simply, it was added mainly as a means of increasing the efficiency of combustion although it was also thought to have other beneficial effects, particularly on valve function. The practice started in the 1920s and continued through the better part of the twentieth century. The amounts added to petrol were not insignificant, averaging around 0.14 g per litre. This means that at least 20–30 g of lead additive would have been released into the environment by filling a small car and running it till it was empty. Of course, as most of us fill our car with petrol many times a year it is easy to work out that average motorist would have released hundreds of grams of lead into the environment every year and that the combined lead release from the many hundreds of thousand of motorists in Australia and other developed countries would have run into the thousands of tonnes. Combustion products containing lead have also been reported to be present in small particles termed "PM" (for particulate matter) the size of thes particles varying in size.

When lead tetraethyl petrol is burnt, lead itself is released and, if left to accumulate would destroy the engine. To get around this problem, other substances are added to "mop up" the lead. The two most common that were used over the years were dichloroethane (DCE) and dibromoethane (DBE). During the combustion process, and in the presence of DCE and DBE, the lead that is released from tetraethyl becomes volatile and released into the atmosphere through the exhaust. The net result of all of this is that the lead in petrol is released into the atmosphere in a volatile form, one that that is easy to breathe into the lungs and, from there, into the bloodstream and throughout the body.

While the principal exposure was as lead products formed from the lead in petrol, smaller amounts of lead are also released, and continue to be released, from many other sources. One source which has generated considerable concern, particularly in the area of public health, is the lead produced from smelting, ie the process that extracts the metal from its ore which is found in the ground. The process itself is fairly straight forward involving the heating of the ore (the most common ore is galena which is composed of a mixture of lead and sulphur) in air. During the process, there is release of sulphur containing gases as well as some lead containing dust and perhaps even volatile forms of lead. Some of the dust finds its way into the atmosphere. It may then deposit locally or, after being carried in air currents, many miles away. Soil analysis of major smelting areas have confirmed that there may be increased levels of lead in the soil [59]. In addition, atmospheric levels of lead are also increased in the vicinity of lead smelters. Occupational exposure to lead, as occurs for example in the battery, plastics, ceramics, glazes, and paint industries to name just a few, is also a very important source of lead exposure.

Another source of exposure is lead in food. The Australian Total Diet Surveys conducted by Australia New Zealand Food Standards, the Australian Government agency charged with monitoring food for contaminants have confirmed that traces of lead occur in almost all of the common foods (19th and 20th Surveys). In earlier years, when lead solders where used in the canning of food, some of the lead was due to its migration of from the solder [2]. It confirms, yet again, that packaging materials including cans are not necessarily inert and may react with food.

It is clear that mining, and the associated smelting, is a very significant contributor to exposure to lead. Perhaps the best known example of this is the small township of Port Pirie, situated in South Australia around 200 km from Adelaide and more recently, Mount Isa in Queensland [42]. Lead smelting commenced in Port Pirie in the late 1800s and by 1934 Port Pirie was the site of the largest lead smelter in the world. However, while lead smelting led to benefits for residents of the town, these benefits were offset somewhat by the gradual realization that these benefits had a cost- the increasing contamination of the environment with lead. Numerous studies over many years have confirmed that both air and soil in parts of the town contain higher than normal levels of lead. As far back as 1986, a survey of the blood of 1239 Port Pirie children (or 50% of the elementary school children) around 1 in 15 of the children had lead levels that were equal to or greater than 30 micrograms per 100 ml, a level considered by the National Health and Medical Research Council (the premier medical research and public health body in Australia) to be of considerable concern because of the known toxic effects of lead [3]. In the following year, another group of researchers found that pregnant women who had resided in Port Pirie for more that 3 years, or who had lived in parts of Port Pirie that had high lead content, had higher than normal levels of lead in their blood [4]. The concern here was that the increased blood lead levels could impact on the health of the developing foetus. Lead exposure associated with mining or smelting is not confined to Port Pirie because there have been many reports of exposure all over the world including other parts of Australia (Mount Isa), Brazil, Thailand, and, more recently and tragically, Nigeria [42–44, 60, 67].

What Are "Safe" Blood Levels of Lead?

In 1991, the US Centres for Disease Control and Prevention determined that a level of 10 microgram per deciliter could be considered safe, a figure that was much lower than the 30 micrograms per decilitre to which many of the Port Pirie children had been exposed. Of course, what this means is that for many years people in Port Pirie were exposed to levels of lead that, in the light of new evidence, were now considered potentially harmful. Moreover, in more recent years, there have even been indications that this lesser figure of 10 micrograms per decilitre is still too high and a figure of 5 micrograms or less is now thought to be even safer [5, 68].

Irrespective of how lead gets into the body, it is carried by the blood and deposited in various organs. Lead has been found in most organs examined including hair, toenails, kidney, brain and the eye [6, 7, 40]. More alarmingly, it has also been found in the brain and kidneys of foetuses as well as in neonates [8].

What Is the Evidence That Lead Is Harmful?

Given that it does find its way into our blood and organs, what is the evidence that it is harmful? There is no doubt that, at high levels, lead is toxic causing abdominal pain, lethargy, anorexia, anemia, problems with walking, and slurred speech. Convulsions, coma, and death can occur if the dose is high enough. Occupational exposure, for example to lead stabilizers used in plastics, or to lead glazes used in pottery, may result in severe poisoning with a number of these symptoms [9, 10]. Severe lead poisoning has also been reported from drinking from a lead glazed mug (the lead from the glaze dissolved in hot tea), from a course of treatment with ayervedic drugs, or even through the swallowing of lead jewellery [11, 12, 13].

The chronic exposure, that is the exposure to small, apparently non toxic amounts, over extended periods of time, is much more widespread, and a potentially greater public health problem although its effects on our health are difficult to quantify. However, there is accumulating evidence the chronic exposure to small, and apparently non-toxic levels of lead, can have effects. The part of our body that appears the most susceptible is the brain, particularly in very young children. The brain continues to develop at birth. One of the most significant changes is the deposition of the myelin sheath, which has been likened to a sort of insulation covering some nerve fibres and greatly assists the transmission of impulses along these fibres. Myelin is critically important because abnormalities in myelin can cause disease –for example multiple sclerosis, a degenerative brain disease is caused by loss of the myelin sheath. Another reason for the increased susceptibility of the brain in young children is the incomplete development of the blood brain barrier, highly specialized blood vessels which prevent the entry of substances that are likely to harm the brain as discussed in Chap. 2. This barrier is not fully formed at birth so potentially toxic substances are more likely to cross into the brain. There is no doubt that lead does cross into the brain because it has been found in the brains of very young children and even in the fetus [8].

There are now numbers of reports on the possible effects of small amounts of lead on brain function in children. Many of these studies have been prompted by concerns for children living in or near lead smelting towns like Port Pirie. The function of the brain that has been most studied in relation to lead exposure is the measurement of the intelligence quotient (IQ) which in itself provides an indication of what is often referred to as cognition or cognitive ability ie the ability of the brain to perceive, think and remember. These are highly complex processes but little understood. While the measurement of IQ is not a perfect measure of the cognitive functioning of the brain, it is easily measured, been in use for many years, and there is a lot of information on what is normal. Using IQ as a general yardstick, there have been numerous studies, some going back a decade or more, that suggest that children exposed to small amounts of lead have lower IQs [14, 15, 16] Whereas in the past, it was considered that levels below 10 microgram/decilitre (a decilitre is literally in one tenth of a litre or 100 ml), there are indications that even levels below this may affect IQ [15, 17]. It is also thought that exposure of pregnant women to lead, particularly around 28 weeks gestational age, can result in the birth of children with lowered IQs [18].

While most of the studies have focused on the effects of lead on children, there is increasing evidence that lead can have effects on the health of adults. For example, there are reports that lead exposure can lead to an atypical type of Parkinson's disease, and motor neurone disease, as well as kidney disease [19, 20, 21]. In the case of kidney disease, it has been suggested that individuals with high blood pressure or diabetes may have an increased risk. There are also suggestions of an association between lead exposure and hypertension (high blood pressure) [22].

Mercury

What Are the Industrial Uses of Mercury?

Mercury occurs quite naturally in nature, mainly as cinnabar or mercuric sulphide (ie mercury and sulphur atoms combined). Trace amounts of mercury occur naturally in rivers, in the ocean, and in the soil. Much of the metal which is present in the environment comes from the combustion of fossil fuels, mining, and industrial activities. Most people associate the use of mercury with thermometers but it has many industrial uses, including the manufacture of batteries, industrial instruments, alloys, fluorescent lights as well as in chlorine-alkali production. It is also released from raw materials (limestone) and fuel (coal) used in cement production. Until recently large amounts of mercury were also used extensively in the manufacture of paints but manufacturers have agreed to discontinue the use of the metal, at least in interior paints.

Mercury in Food

Given that mercury is known to be toxic, it is difficult to understand how it could be added deliberately to food but this is what happened in the early 70s when a fungicide containing mercury was used to treat grain in Iraq. The mercury-treated grain was used to make bread which led to the exposure of thousands of people and the death of hundreds. Some of the effects in children included mental retardation, blindness, deafness, and delayed development [41].

The realization that mercury from industrial wastes could enter the food chain inadvertently stems largely from reports of the poisoning of people living in Minamata City, in the southwest region of Kyushu Island of Japan in the 1950s. The source of mercury was later shown to be fish from Minamata Bay [23]. Since that time, it has become very clear that the contamination of fish with methylmercury is not confined to the seas around Japan alone as significant, although much smaller levels, have been found in fish in other parts of the world [24, 25]. While mercury is present in very small amounts in smaller fish, its levels are much higher in larger fish such swordfish, tuna and other predator fish such as shark, and marlin [26, 27]. It is believed that the larger concentrations of mercury in the bigger fish are due to what is termed 'biomagnification' of the metal ie the very small fish containing mercury are eaten by larger fish thereby increasing the level of mercury in their tissues and they in turn are consumed by still larger fish and so on up the food chain.

What Are the Concerns About Mercury?

Mercury is known to affect the function of many organs of the body including the brain and the kidneys. Acute exposure of the metal (ie exposure to high doses) may lead to damage of the intestine, heart, blood vessels, and kidneys and can be fatal. Chronic exposure (ie exposure to smaller, non-fatal amounts) can occur in different ways – in certain foods, through the use of vaccines containing the preservative thimoresal, through the use of skin creams and infant teething powders containing mercury, from dental amalgams, occupational exposure and accidents, and even complementary medicines (Ayurvedic), and skin care products [46]. However, for most people, the most important source is food. For very young children, their principal source of food ie breast milk has also been found to contain the metal [61]. The burning of coal is another source of mercury [63]. Traces of mercury are found in many foods, but the greatest dietary exposure occurs through the consumption of contaminated fish [50]. Much of the mercury in fish and other foods is present in an organic form (ie mercury combined with carbon and hydrogen). Organic forms of mercury (mainly methylmercury) are derived from microorganisms and are also toxic. Another source of exposure is through dental amalgams.

The effects of consuming methylmercury contaminated fish on humans are well known from the outbreak of Minamata disease in Japan mentioned above as well as from the Iraq poisoning. Minamata disease resulted from the discharge of mercury from a chemical plant into Minamata Bay and its subsequent uptake by marine life which were in turn eaten by people who lived in the area. The marine products were shown to have very high levels of mercury, from around 5, to in some cases reaching levels of up to 35, parts per million [23]. Typical symptoms mostly related to effects on the brain and included abnormalities in walking, speech, hearing, and vision. Tremor was also a common feature. Consumption of contaminated fish also caused birth defects, particularly affecting the brain, indicating that the methylmercury had passed from the mother to the unborn child [62].

The effects of exposure to much smaller amounts of mercury are not so clear cut. However, there are indications that occupational exposure of workers as occurs for example in chloro-alkali plants, thermometer manufacture, the chemical industry, and mining, can affect coordination [28, 47]. Low level exposure from food, especially fish, and other sources, as determined by measuring the levels of mercury in hair, has been reported to increase the likelihood of tremor in adults [29, 48]. The Minemata outbreak confirmed that the very young and the unborn are particularly susceptible to low level exposure of mercury because their brains are not fully developed [62]. However, recent studies indicate that while high seafood consumption is associated with a greater intake of mercury, there is no evidence that the health of the fetus is compromised [52, 71]. These findings, together together with those from the Minemata poisoning, suggest that mercury levels in fish have to be above a certain, as yet undetermined level, before there are any measurable effects on health. Despite this, governments have become so concerned that they have issued warnings to pregnant women wanting to become pregnant, as well as mothers of small children, to limit their intake of certain fish [30, 31]. These warnings have been prompted at least in part by the demonstration that some women of child-bearing age have blood mercury levels in excess of the recommended levels.

There is some evidence that exposure to the small amounts of mercury in dental amalgams may produce minor changes in kidney function but there is nothing to indicate that these changes lead to disease [53].

Arsenic

What Are the Industrial Uses of Arsenic?

Arsenic is a naturally occurring element found mostly in the combination with other elements such as oxygen and sulphur. One of its chief uses is as a wood preservative but it also used in metallurgy, for semiconductor materials in the electronics industry, and in glass manufacture. It has been used to kill parasites in veterinary medicine, and also for homeopathic and folk remedies in different parts of the world. Until the early 1940s it was also used to treat syphilis and psoriasis.

Arsenic in Ground Water and Coal

Because of its presence in the earth's crust, some arsenic finds its way into our drinking water. While in most parts of the world, the amounts are very small, rarely exceeding 10 micrograms (ie parts per million or millionths of a gram) or 50 micrograms per litre, the maximum permissible amount for drinking water recommended by the World Health Organisation or the American Council on Science and Health respectively. In some parts of West Bengal in India, and Bangladesh, there may be as much as 3000 micrograms per litre [32, 33]. There is clearly some disagreement as to what the acceptable levels of the metal in drinking water are. Coal is another source of arsenic. It has been estimated that there may be as much as 3 mg per kg of arsenic in some coal and burning may release many tonnes of the metal into the environment every year [63].

Arsenic in Food

Because of its widespread occurrence, it is not unexpected that arsenic finds its way into our food [50]. According to the figures quoted by the FSANZ 23rd Total Diet Survey, the diet of an average 70 kg male contains up to nearly 100 micrograms per day although there have been reports in other parts of the world of uptakes as high as 223 micrograms per day in males [34] Seafood is a major source of arsenic in the diet with some fish, notably shark, containing up to 14 micrograms per gram which means that a single 100 g serve would result in an intake of as much as 1400 micrograms (or 20 times the normal 70 kg male intake). Rice is another source of arsenic and while the levels may not be as high as in some seafood, it is present in the more toxic inorganic form and therefore of some concern particularly for young children [66, 69]. The amounts in food also appear to be influenced by the levels of arsenic in water because vegetables grown in areas like Bangladesh which have contaminated ground water have considerably more arsenic than those grown in other countries [35]. In some parts of the world, organic forms of arsenic are added to poultry feed and find their way into the food chain [65].

What Are the Concerns About Arsenic?

Like mercury, arsenic exists in two forms – organic and inorganic. Whereas the inorganic form predominates in water, arsenic which is in the food chain is mainly in the organic form ie it is found in combination with carbon and derived from living organisms. The devastating consequences of contamination of food with arsenic was reported in the mid 1950s in Japan when 130 children died as a consequence of ingesting arsenic contaminated milk powder. The actual number who died was later shown to be in excess of 600. The source of the toxic metal was sodium phosphate, a rather innocuous chemical added to milk powder but contaminated with inorganic arsenic. It is well known that exposure to large amounts of inorganic arsenic can lead to difficulty swallowing, nausea, vomiting, abdominal pain, diarrhea, and death. Chronic exposure, ie exposure to smaller amounts (more than a few hundred millionths of a gram per litre of water), which is a particular problem in parts of Bangladesh and India, over a period of time, can give rise to high blood pressure, skin abnormalities, effects on the heart and blood vessels, and may cause cancers of the skin, liver and prostate [33, 64]. Chronic exposure to smaller amounts is also believed to increase the risk of death from cardiovascular (strokes, heart disease) and kidney disease and the prevalence of kidney, eye and nerve abnormalities in diabetics [36, 51, 52]. There are suggestions as well of an increase in spontaneous loss of pregnancy [49]. Whether exposure to small amounts of arsenic is harmful is thought to depend on genetic factors [64]. Very young children appear to be particularly susceptible because many of the Japanese children exposed to arsenic contaminated milk powder were later shown to have mental retardation and their growth was impaired [69, 70]. It is believed that a total dose of 60 mg (60 thousandths of a gram) of arsenic over this critical developmental period was sufficient to affect growth and development [70].

Organic arsenic has not thought to be especially harmful but there are doubts about this proposition because of the diversity of organic arsenic substances which exist in living things, including the plants and animals we eat, and for which there are limited safety data [37]. There is now clear evidence that at least one form of organic arsenic, dimethylarsenic acid or DMA, which is produced quite naturally in the body from the inorganic form, has potent carcinogenic properties, at least in rats and mice [38]. Indeed, it has been proposed that the reason inorganic arsenic is able to increase the risk of cancer and other diseases is because it is converted by the body into certain organic forms, for example DMA, which in turn can react with, and affect the function of, key proteins and the DNA in our genes [39]. If this theory is correct, then the currently accepted view that all organic forms of arsenic are non toxic has to be re-evaluated. As the food that we eat is likely to contain many different organic forms, it would be indeed surprising if some of these other substances were not as harmful as DMA.

What Does All of This Mean?

The toxicity of lead, arsenic, and mercury has been known for a long time. It is therefore truly astounding that governments around the world permitted the use of lead in petrol when it was clear that, with an ever increasing number of petrol driven motor vehicles on our roads, the levels of lead entering the atmosphere, particularly in heavily populated urban areas, would also increase. It is difficult to understand the logic of releasing a known neurotoxin into the environment, thereby exposing entire communities to the metal. Similarly, it is incomprehensible, at least with knowledge we have today, that mercury, another neurotoxin was incorporated into vaccines for children and amalgams used as fillings for teeth, and arsenic, another toxic metal was used as a preservative for wood used in children's playgrounds. It also not clear whether any thought was given to how the environment would cope with this added toxic burden. Unlike many other pollutants that are released into the environment which can be degraded by microbes into simpler, less toxic substances, toxic metals such as lead, mercury and arsenic, are stable and remain in the environment in forms that retain a level of toxicity. In a way they are like the organochlorines which are similarly stable but, unlike the organochlorines which do eventually break down over a period of many years, toxic metals remain in our soils and water in some form or other forever. Of course, governments may argue that the amounts that are released are not toxic but the Minnemata incident demonstrated that even amounts that are, in themselves, non toxic can, through the processes of biomagnification in fish and other animals, build up to levels that are potentially harmful to humans. Moreover, if governments really believed that toxic metals like mercury are non hazardous in the amounts we are exposed to, why then are there governmental warnings to pregnant women to limit their intake of the larger fish? Also, while the amounts we are exposed to are small, and at these levels, apparently non-toxic, there is ample proof that, in the same way that the larger predator fish accumulate these metals in their tissues through their food, humans also seem to be able to do the same thing. While the amounts of the toxic metals in the whole organs of humans are small, there is the possibility, rarely considered, that their levels may be sufficiently high in certain specialized parts of organs to affect the function of the whole organ. The fact is that these metals have already caused harm to communities around the world. Unless drastic measures are taken to limit their use or, if this is not practical, more stringent controls are put in place to reduce their release into the environment, they will continue to do so.

References

1. Clark CS et al (2006) The lead content of currently available new residential paint in several Asian countries. Environ Res 102:9–12
2. Meah MN et al (1991) Lead and tin in canned foods: results of the UK survey 1983-1987. Food Addit Contam 8:485–496

3. Wilson D et al (1986) Children's blood lead levels in the lead smelting town of Port Pirie, South Australia. Arch Environ Health 41:245–250
4. Baghurst P et al (1987) Determinants of blood lead concentration of pregnant women living in Port Pirie and surrounding areas. Med J Aust 146:69–73
5. Gilbert SG, Weiss B (2006) A rationale for lowering blood lead action level from 10 to 2 microg/dL. Neurotoxicology 27:693–701
6. Anwar M (2005) Arsenic, cadmium, and lead levels in hair and toenail samples in Pakistan. Environ Sci 12:71–86
7. Barregard L et al (1999) Cadmium, mercury, lead in kidney cortex of the general Swedish population: a study of biopsies from living kidney donors. Environ Health Perspect 107:867–871
8. Lutz E (1996) Concentrations of mercury, cadmium, and lead in the brain and kidney of second trimester fetuses and infants. J Trace Elem Med Biol 10:61–67
9. Coyle P et al (2005) Severe lead poisoning in the plastics industry: a report of three cases. Am J Ind Med 47:172–175
10. Shiri R et al (2007) Lead poisoning and recurrent abdominal pain. Ind Health 45:494–496
11. Kirchgatterer A et al (2005) Weight loss, abdominal pain and anemia after a holiday abroad – case report of lead poisoning. Dtsch Med Wochenschr 130:2253–2256
12. Schulling U et al (2004) Lead poisoning after ingestion of ayervedic drugs. Med Klin (Munich) 99:476–480
13. Weldenhamer JD (2007) Widespread lead contamination of imported low cost jewelery in the US. Chemosphere 67:961–965
14. Tong S et al (1996) Lifetime exposure to environmental lead and children's intelligence at 11-13 years: the Port Pirie cohort study. BMJ 312:1569–1575
15. Canfield RL et al (2003) Intellectual impairment in children with blood lead concentrations below 10 microgram per decilitre. N Engl J Med 348:1517–1526
16. Surkan PJ et al (2007) Neuropsychological function in children with blood lead levels < 10 migrog/dL. Neurotoxicology 28:1170–1177
17. Jusko TA et al (2008) Blood lead concentrations <10 mug/dL and child intelligence at 6 years of age. Environ Health Perspect 116:243–248
18. Schnaas L et al (2006) Reduced intellectual development in children with prenatal lead exposure. Environ Health Perspect 114:797–797
19. Sanz P et al (2007) Progressive supranuclear palsy-like parkinsonism resulting from occupational exposure to lead sulphate batteries. Int Med Res 35:159–163
20. Fang F et al (2010) Association between blood lead and the risk of amyotrophic lateral sclerosis. Am J Epidemiol 171:1125–1133
21. Ekong EB et al (2006) Lead related nephrotoxicity: a review of the epidemiological evidence. Kidney Int 70:2074–2208
22. Navas-Acien A et al (2007) Lead exposure and cardiovascular disease – a systematic review. Environ Health Perspect 115:472–482
23. Harada M (1995) Minamata disease: methylmercury poisoning in Japan caused by environmental pollution. Crit Rev Toxicol 25:1–24
24. Cabanero AI et al (2005) Quantification and speciation of mercury and selenium in fish samples of high consumption in Spain and Portugal. Biol Trace Elem Res 103:17–35
25. Bjornberg KA et al (2005) Methyl mercury exposure in Swedish women with high fish consumption. Sci Total Environ 341:45–52
26. Storelli et al (2004) Content of mercury and cadmium in fish (Thunnus alalunga) and cephalods (Eledone moschata) from the south-eastern Mediterranean Sea. Food Addit Contam 21:105–1056
27. Fossi et al (2004) Evaluation of ecotoxicological effects of endocrine disrupters during a four-year-survey of the Mediterranean population of swordfish (Xiphias gladius). Mar Environ Res 58:425–429

28. Lucchini R et al (2002) Neurotoxic effects of exposure to low doses of mercury. Med Lav 93:202–214

29. Auger et al (2005) Low level methylmercury exposure as a risk factor for neurologic abnormalities in adults. Neurotoxicology 26:149–157

30. Food Standards Australian New Zealand. www.foodstandards.gov.au

31. Centers for Disease Control and Prevention (2005) Blood mercury levels in young children and child-bearing women- United States, 1999–2002. MMWR Morb Wkly Rep 53:1018–1020

32. Bhattacharyya R et al (2003) High arsenic groundwater: mobilization, metabolism, and mitigation – an overview in the Bengal Delta plain. Mol Cell Biochem 253:347–355

33. Brown KG et al (2002) Arsenic, drinking water, and health: a position paper of the American Council on Science and Health. Regul Toxicol Pharmacol 36(162–174):2002

34. Llobet JM et al (2003) Concentrations of arsenic, cadmium, mercury, and lead in common foods and estimated daily intake by children, adolescents, adults, and seniors of Catalonia, Spain. J Agric Food Chem 51:838–842

35. Al Rmalli et al (2005) SW. A survey of arsenic in foodstuffs on sale in the United Kingdom and imported from Bangladesh. Sci Total Environ 337:23–30

36. Chiou JM et al (2005) Arsenic ingestion and increased microvascular disease risk: observations from the south-western arseniasis-endemic area in Taiwan. Int J Epidemiol

37. Jomova K et al (2011) Arsenic toxicity, oxidative stress and human disease. J Appl Toxicol 31:95–107. http://www.ncbi.nlm.nih.gov/pubmed/21321970

38. Wanibuchi H et al (2004) Understanding arsenic carcinogenicity by use of animal model. Toxicol Appl Pharmacol 198(366–376):2004. http://www.ncbi.nlm.nih.gov/pubmed/15276416

39. Kitchin KT (2001) Recent advances in arsenic carcinogenesis:modes of action, animal model systems, and methylated arsenic metabolite. Toxicol Appl Pharmacol 172(249–261):2001. http://www.ncbi.nlm.nih.gov/pubmed/11312654

40. Eichenbaum JW, Zheng W (2000) Distribution of lead and transthyretin in human eyes. J Toxicol Clin Toxicol 38:377–381

41. Bakir F et al (1980) Clinical and epidemiological aspects of methymercury poisoning. Postgrad Med 56:1–10

42. Taylor MP, Schniering C (2010) The public minimization of the risks associated with environmental lead exposure and elevated blood lead levels in children, Mount Isa, Queensland, Australia. Arch Environ Occup Health 65:45–48

43. Paoliello MM et al (2002) Exposure of children to lead and cadmium from a mining area of Brazil. Environ Res 88:120–128

44. Pusapukdepob J et al (2007) Health risk assessment of villagers who live near a lead mining area: a case study of the Klity village, Kanchanaburi Province, Thailand. Southeast Asian J Trop Med Public Health 38:168–177

45. BBC News. www.bbc.co.uk. 4 June 2010

46. McKelvey W et al (2010) Population-based inorganic mercury biomonitoring and the identification of skin care product as a source of exposure in New York City. Environ Health Perspec

47. Bose-O'Reilly S et al (2010) Health assessment of artisanal gold miners in Indonesia. Sci Total Environ 408:713–725

48. Risher JF (2004) Too much of a good thing (fish): methylmercury case study. J Environ Health 67:9–14

49. Bloom MS et al (2010) Spontaneous pregnancy loss in humans and exposure to arsenic in drinking water. Int J Hyg Environ Health

50. UK total diet survey (2006) Measurement of the concentration of metals and other elements from the UK total diet survey. http://cot.food.gov.uk/pdfs/cotstatementtds200808.pdf

51. Moon K et al (2012) Arsenic exposure and cardiovascular disease: an updated systematic review. Curr Atheroscler Rep 14:542–545. http://www.ncbi.nlm.nih.gov/pubmed/22968315

52. Meliker JR et al (2007) Arsenic in drinking water and cerebrovascular disease, diabetes mellitus, and kidney disease in Michigan: a standardized mortality ration analysis. Environ Health 6:4. http://www.ncbi.nlm.nih.gov/pubmed/17274811

53. Geier D et al (2012) A significant dose-dependent relationship between mercury exposure from dental amalgams and kidney integrity marker: a further assessment of the Casa Pia children's dental amalgam trial. Hum Exp Toxicol. Epub. http://www.ncbi.nlm.nih.gov/pubmed/22893351

54. Li HB et al (2012) Lead contamination and source in Shanghai in the past century using dated sediment cores from urban park lakes. Chemosphere 88:1161–1169. http://www.ncbi.nlm.nih.gov/pubmed/22537888

55. Zhang L et al (2012) Influence of mercury and chlorine content of coal in the past century using dated sediment cores from urban park lakes. Environ Sci Technol 46:6385–6392. http://www.ncbi.nlm.nih.gov/pubmed/22533359

56. Kang Y et al (2012) Arsenic in Chinese coals: distribution, modes of occurrence, and environmental effects. Sci Total Environ. http://www.ncbi.nlm.nih.gov/pubmed/22078371

57. http://www.epa.gov/mats/actions.html

58. http://www.epa.gov/wastes/nonhaz/industrial/special/fossil/ccr-rule/index.htm

59. Li P et al (2015) Contamination and health risks of soil heavy metals around a lead/zinc smelter in southwestern China. Ecotoxicol Environ Saf 113:391–399

60. Greig J et al (2014) Association of blood lead level with neurological features in 972 children affected by and acute severe lead poisoning outbreak in Zamfara State, Northern Nigeria. PLoS One 9

61. Iwai-Shimada M et al (2015) Methylmercury in breast milk of Japanese mothers and lactational exposure of their infants. Chemosphere 126:67–72

62. Grandjean P, Herz KT (2011) Methylmercury and brain development: imprecision and underestimation of developmental neurotoxicity in humans. Mt Sinai J Med 78:107–118

63. Kang Y et al (2011) Arsenic in Chinese coals: distribution, modes of occurrence, and environmental effects. 412-413, 1-13

64. Karagas MR et al (2015) Drinking water contamination, skin lesions, and malignancies: a systematic review of the global evidence. Curr Environ Health Rep 2:52–68

65. Fisher DJ et al (2015) Environmental concerns of roxarsone in broiler poultry and litter in Maryland, USA. Environ Sci Technol 49:1999–2012

66. Hojsak I et al (2014) Arsenic in rice – a cause for concern. J Pediatr Gastroenterol Nutr

67. Dong C et al (2015) Environmental contamination in an Australian mining community and potential influences on early childhood health and behavioural outcomes. Environ Pollut 207:345–356

68. NHMRC Statement and Information Paper: Evidence on the effects of Lead on Human Health (2016). https://www.nhmrc.gov.au/guidelines-publications/eh58

69. Davis MA et al (2017) Assessment of human dietary exposure to arsenic through rice. Sci Total Environ 586:1237–1244. 69. Yorifuji T et al (2017) Height and blood chemistry of adults with a history of developmental arsenic poisoning from contaminated milk powder. Environ Res 155, 86–91

70. Dakeishi M et al (2006) Long term consequences of arsenic poisoning during infancy due to contaminated milk powder. Environ Health 5:31

71. Davidson PW et al (2011) Fish consumption and prenatal methylmercury exposure: cognitive and behavioural outcomes in the main cohort at 17 years from the Seychelles child development study. Neurotoxicol 32, 711–716. http://www.ncbi.nlm.nih.gov/pubmed/21889535

Chapter 5
The Indestructibles

The amounts, number, and variety, of chemicals used by industry and the waste products derived them is staggering. For example, the US Environmental Protection Authority (EPA) has around 80,000 different chemicals on its chemical inventory. Its website lists some of the industries e.g. construction, dry cleaning, laboratories, motor freight, furniture manufacturing etc., as well as chemicals associated with the various activities [71]. The chemicals include liquids, gases or volatiles, and solids. There are too many chemicals to examine in any detail. In this chapter it is proposed to discuss three classes of these chemicals, both of which have been found to break down very slowly and hence have significant environmental stability. These include the polychlorinated biphenyls (PCBs), dioxins, and the polybrominated biphenyls (PBDEs). Two other classes of environmentally stable chemicals, the organochlorine pesticides and fluorocarbons, have been discussed in earlier and later chapters of this book respectively. But before we do this, it is worth looking at the regulation of industrial chemicals in some parts of the world.

Regulation of Industrial Chemicals

The EPA is involved in the assessment of the environmental and human health risks of chemicals used in the USA. The corresponding body in Europe is the European Chemicals Agency (ECA) and the National Industrial Chemical Notification Assessment Scheme (NICNAS) was established by the government to assess the environmental and human health risks of chemicals used in Australia. There has been some criticism of role played by these agencies in the regulation process because of the reliance the regulatory authorities place on information obtained from the industrial users about the likely hazards of a particular chemical. It has been argued that any information relating to an individual chemical provided by the user may be biased and therefore not subjected to the same scrutiny as required for publication in a peer reviewed scientific journal. As the number of chemicals already registered is large and increasing every year, and as the resources of government agencies are limited, it is not surprising that some chemicals quite literally have "fallen through the cracks" and are banned many years after approval.

Polychlorinated Biphenyls (PCBs)

The term "polychlorinated biphenyls" or PCBs actually describes a great number of different substances, made up of carbon, hydrogen and chlorine, which can differ chemically in many ways but in particular differ according to the proportion of chlorine relative to carbon atoms. They were first manufactured in the early part of the twentieth century, and were used in many different products including hydraulic fluid, pigments, carbonless copy paper, plasticizers, vacuum pumps, compressors, and also in electrical equipment, in particular transformers and capacitors. They are chemically very stable, resistant to heat and like the organochlorine pesticides, and have a high affinity for fat. Unfortunately, while their chemical stability was considered an asset early on, this was later shown to be a bit of a two -edged sword because of their remarkable stability when released into the environment. This same property is shared with the organochlorine pesticides. This environmental stability, coupled with their widespread use, led eventually to a gradual accumulation in rivers, lakes, the ocean, soil and even the polar ice cap. The next step in this process, their accumulation in marine and freshwater animals, followed closely by their uptake into the food chain was, in hindsight, not surprising and shows once again that the consequences of environmental release of chemicals is not often understood at the time of initial release.

PCBs are relatively stable to heat, but at higher temperatures (in excess of 200 °C) and in the presence of oxygen, their chemical structure begins to change and they form substances that are defined chemically as "dioxins". The best known of the dioxins formed under these conditions is PCDD. Another group of substances, called "PCDFs" (defined chemically as "furans") are also formed from PCBs [1]. However, if the temperature goes beyond about 800°, the "dioxins" and "furans" are eventually broken down completely into much simpler substances such as carbon dioxide. If electrical or other equipment, or other goods containing PCBs, are burnt in municipal incinerators, particularly if high temperatures are not used, or in the event of a fire, dioxins may be formed in significant amounts and released into the atmosphere from where they may then deposit on land or sea. Small amounts of PCDFs, are also present as contaminants in PCBs and released into the environment together with the PCBs. Like the PCBs which they resemble chemically, dioxins are stable with an affinity for fat.

Do PCBs and Dioxins Enter the Food Chain?

The answer to this question is most definitely yes. Analyses of foods from different countries around the world have confirmed the presence of PCBs, together with PCDFs and PCDDs in food although the amounts present in most cases are very small [2–4, 41, 67, 68], Fish and shellfish are a major source and the amounts in the larger predator fish, such as for example, may be relatively large [2–5]. There are indications that farmed fish may sometimes contain more than wild fish [6].

Are the PCBs and Dioxins Taken up into the Body?

There is now a wealth of information demonstrating unequivocally that PCBs and dioxins present in food do end up in our bodies. Food is not the only source of exposure to these substances because they have also been shown to be present in both indoor and outdoor air and can therefore enter the body via inhalation [66]. Most of the studies have measured levels in blood but PCBs have also been found in other tissues such as fat, breast milk, placenta (the organ joining a mother to her unborn baby), muscle, and even lungs [7–9, 52]. The amount taken up is related to the levels in food, which is why people who have a high fish diet, particularly when the fish and other seafood is contaminated with PCBs and dioxins may have higher than normal amounts of these substances in their bodies [10–12, 42]. Of perhaps even greater concern is the presence of PCBs and dioxins in breast milk. There are numerous reports from around the world of contamination of breast milk with PCBs and dioxins [10, 13, 14]. Breast milk of women from the Faroe Islands, an archipelago of islands lying in the north Atlantic Ocean, was shown to contain about 2 µg (millionths of a gram) per gram of milk fat [10]. Assuming that a baby takes in about 700 ml of breast milk containing 4% fat i.e. 28 g of fat per day, the amount of PCBs ingested is about 50–60 µg per day, a very significant amount when related back to the weight of a new born baby.

What Do PCBs and Dioxins Do to Animals?

Studies with animals have confirmed that PCBs and dioxins can induce cancer, and affect bone, the thyroid gland, and sexual development [43, 58–60]. While these effects are mostly produced by amounts much greater than those to which humans are exposed, nevertheless it does raise quite genuine concerns about the impact these substances may be having on humans [54].

Can PCBs and Dioxins Affect Our Health?

In 1968, accidental poisoning (referred to as the Yusho incident), later shown to be due to contamination of rice oil with PCBs, occurred in Northern Kyushu, Japan, followed about 10 years later with similar mass poisoning (also with contaminated oil) in central Taiwan (termed Yucheng meaning oil disease in Chinese). The principal symptoms included a severe type of acne (termed "chloroacne"), increased pigmentation of the skin, and eye discharge. The effects of PCB exposure in Japan and Taiwan were not confined to adults. Children born to mothers exposed to PCBs while pregnant or shortly after birth showed a higher than normal incidence of growth retardation, birth defects, ear inflammation, behavioural problems, and abnormal tooth development [18, 19].

The accidental poisoning that occurred in Japan and Taiwan resulted in exposure to relatively large amounts of PCBs and PCDFs and perhaps does not really reflect the more usual smaller exposure. There is no doubt that occupational exposure, for example such as occurred in a transformer recycling plant, does lead to increases in blood levels of PCBs [52]. Diet is also an important source of PCBs, particularly seafood [53]. One group with a relatively high exposure to dietary PCBs, but not on the same scale as either Yusho or Yucheng, is the Akwesasne Mohawks who live in an area spanning the St Lawrence river and the boundaries of New York State, Ontario, and Quebec. This group has been exposed to amounts of PCBs in excess of the maximum amounts recommended by the USA Food and Drug Administration.

Analysis of PCBs in the blood of over 100 10–16 year old girls demonstrated a relationship between the levels of PCBs and the age at which the girls commenced menstruation. The studies showed that the higher the blood levels of PCBs, the greater the likelihood that menstruation had occurred earlier and indicated that the PCBs had, in some way, interfered with the normal hormonal processes that regulated menstruation [12]. There is evidence that PCBs may be able to do this by mimicking the effects of natural hormones in the body.

There are suggestions that environmental exposure to PCBs can increase the risk of certain cancers such as breast and non-Hodgkins lymphoma but this has been disputed [15, 16, 51]. There is a recent report of a possible link between PCB exposure at work and two diseases affecting brain function i.e. Parkinson's disease and motor neuron disease (caused by the death of nerve cells controlling voluntary movement), and there are suggestions as well that occupational exposure can cause skin abnormalities [17, 50].

The thyroid is a small gland that is situated in the neck and its principal function is to produce a number of hormones including thyroxine, trioodothryronine (often referred to as T3), and calcitonin. These hormones have many functions including effects on blood circulation, appetite, body temperature, to name just a few. Abnormal production (i.e. increased or decreased) of these hormones can cause disease, and deficiencies in the neonatal period can cause mental retardation. Experiments with rats have demonstrated that PCBs can reduce the blood levels of one of the hormones, thyroxine [43]. Effects of PCBs on humans are much more difficult to establish. While there are reports of reductions in the blood levels of thyroxine in individuals whose blood levels of PCBs are increased, this has been disputed [21, 44, 45, 64].

Another group of hormones that have been thought to be affected by PCBs are those involved in male reproduction, and in particular the development of spermatozoa in the testis. There is some evidence that PCBs can affect the function of the testis because semen quality i.e. both numbers of sperm and their movement, and levels of the key male sex hormone, testosterone, are depressed in men with higher blood PCBs but, again, this has been disputed [20, 46].

Perhaps of greater concern are the studies indicating that unborn babies may be particularly sensitive to PCBs even though the exposure of mothers need not be a great as that which occurred during the Yucheng and Yusho contaminations [47]. This view has received support from the findings of a group of American researchers who observed that a higher exposure of pregnant women to PCBs during pregnancy appeared to be associated with a reduced birth weight of their babies as well as a reduction in the circumference of the baby's head [22, 47, 55]. However, once again, a more recent systematic review of a great number of studies has concluded that birth weight is not affected by PCBs. What is alarming, however, is the suggestion that babies exposed before birth to certain PCBs – remember that PCBs are a family of chemical substances – perform badly on the Bayley scales of infant development [23]. The latter measures brain function and includes assessments of language, physical development, and intelligence [56]. As mentioned above, babies are also exposed to PCBs through breast milk which may contain significant amounts of PCBs [10]. There are indications that the combination of pre- and postnatal exposure (through breast milk) may lead to effects on physical development even at 2 years of age [57]. There are even reports that the impact of PCB exposure before birth can be felt in adults through an increased risk of cancer of the testis [24]. Although there have been other reports of an association between different cancers and exposure to PCBs, a recent comprehensive review of the published literature has disputed this [48, 49].

Dioxins are considered to be even more toxic than PCBs. Animal studies have confirmed that even in very small doses, (fractions of millionths of a gram per kilogram of body weight) some dioxins can affect the levels of thyroid hormones, and increase the risk of liver abnormalities and certain cancers (e.g. cancers of the bile duct, liver, uterus, and mouth) [27]. There are suggestions as well that dioxins may interfere with sexual development in animals [28].

Effects of dioxins on humans are more difficult to establish. However, there have been reports of an increased incidence of sarcoma (connective tissue cancers) and non-Hodgkin lymphoma (cancer of the lymph glands) in areas close to municipal solid waste incinerators with high emission levels of dioxins and apparent effects on the sexual development in girls [25, 26, 29].

PCBs, Dioxins and Health – The Bottom Line

It will be very obvious that there is a very great degree of inconsistency in the research into the possible effects of PCBs and dioxins on our health and wellbeing. As soon as one paper is published showing an effect, a year or so later there are others showing that there is no effect. It is therefore very difficult to draw any meaningful conclusions. However, despite this, what we do know is that PCBs and dioxins do enter the food chain, from there they enter our bodies. There is now plenty of evidence that small amounts in animals can produce effects on a number of organs including the brain. There is some suggestive evidence that exposure of either the fetus or very young babies may affect either physical or mental development (or both) but, unfortunately, there is no absolute proof. The jury is still out on this and we await the results of further research.

Polybrominated Diphenyl Ethers (PBDEs)

The polybrominated diphenyl ethers (abbreviated as PBDEs) bear some resemblance to the PCBs but differ from them by containing the element bromine rather than chlorine. There are a number of PBDEs, mostly varying according to the proportion of bromine. They were first developed in the 1970s as fire retardants and used for this purpose in plastic and foam products. Their main uses are in mattresses, bedding, upholstered furniture, building materials, televisions, computers, and other electronic equipment. They are released gradually into the environment where, like the PCBs, they are relatively stable. Once again, and even without the benefit of hindsight, it could have been predicted that their levels in the environment would increase gradually in a manner all too reminiscent of PCBs and organochlorine pesticides. Interestingly, while the levels of PBDEs in the environment are rising rapidly, the corresponding levels of PCBs and organochlorine pesticides have fallen because their use has gradually been phased out. The use of some of the PBDEs has also been discontinued in some parts of the developed world, most notably Europe, and there are indications that levels have begun to drop as a consequence. However, in the USA which has not acted as promptly in banning their use, it is likely that their levels will continue to rise.

Do PBDEs Enter the Food Chain?

There is little doubt that food is a source of much our exposure to PBDEs although there is considerable debate as to whether they can enter the body through other means for example, via house dust [30]. It has been speculated that the ingestion of PBDE-contaminated house dust may be a significant source of PBDEs in young children [30]. There are numerous reports confirming the presence of PBDEs in food. Analysis of US food has shown that fish contains the greatest amounts of PBDEs, followed by meat then dairy [31]. Some fish species, notably sharks, have particularly high levels of PBDEs and, based on analyses made over a 10 year period, there are indications that the levels are rising rapidly perhaps through increasing environmental contamination and gradual build up in tissues [32]. The contamination of food with PBDEs is a general phenomenon which is not confined to the US because PBDEs have also been demonstrated in food from other countries although the levels in the US seem greater than elsewhere [33].

As has been shown for PCBs, PBDEs have also found their way into breast milk [10, 69]. Generally, the amounts are much smaller than the PCBs but it is likely that the amounts will rise with increasing contamination of the environment while PCB levels will fall.

Do PBDEs Get into Our Bodies?

There is no question that PBDEs are taken up into our bodies. Most studies have focused on measuring levels in blood, but PBDEs have also been found in breast milk and body fat [10, 34]. They have been detected in a variety of animal and human tissues [35, 61, 63].

What Do PBDEs Do to Animals?

There are indications PBDEs can affect the development of the brain of rats and mice. Exposed animals have been reported to develop hyperactivity, and there is an impairment in learning and memory tests [36]. There are also indications of reductions in the amounts of the thyroid hormone, thyroxine (also referred to as T4 and contains four atoms of iodine as compared to three for T3, another thyroid hormone) as well as pathological changes in the thyroids of exposed animals [37]. The thyroid is involved in the regulation of many of the chemical reactions taking place in animals, as well as humans, and is particularly important during brain development both before and immediately after birth. Reductions in the amounts of thyroxine formed by the thyroid gland in this early period can result in brain damage in humans and is the reason why the blood of newborn children throughout the developed is tested to ensure that there is no impairment of thyroid function during this critical period. In a recent study, fish exposed to PBDEs were shown to have evidence of liver, kidney and brain abnormalities which indicates that there is an effect on many organs in animals [38]. Impairment of spermatozogenesis (production of sperm) and effects on female reproductive organs have also been described [70].

Can PBDEs Affect Our Health?

Little is really known of the effects of our exposure to the small amounts of PBDEs that are present in the environment. We do know something of the effects of larger amounts of the related substances, polybrominated biphenyls (PBBs) through the accidental contamination of the food chain which occurred in Michigan around 30 years ago. Cattle feed on a number of farms in Michigan became contaminated with PBBs and this resulted in exposure of an estimated 85% of the population of the state. While it is believed that most people were exposed to relatively small amounts, the exposure for some was fairly significant, perhaps as much as 10 g i.e. the equivalent of a couple of teaspoons. Cattle that were fed contaminated feed became emaciated and anorexic, developed skin abnormalities and kidney damage [39]. There have been suggestions that the function of the immune system in some of the people exposed to PBBs in Michigan was reduced and there have been reports of an increased incidence of skin and muscle problems [65].

In the case of PBDEs however, there has been much speculation although little persuasive evidence of a harmful effect on the general population. In a population exposed to PBDEs, for example workers involved in the dismantling of electronic equipment, i.e. residents from a Chinese village close to a factory that dismantled electronic equipment had significantly increased blood levels of PBDEs and, at the same time, higher than normal levels of TSH [40]. TSH is a protein substance produced by the brain which is involved in the regulation of the function of the thyroid. An increase in TSH in the blood is an indication of some sort of change in thyroid function. A similar change has also been found in pregnant women in California [62]. However, no change in the levels of TSH was detected in very young babies even though their mother's milk contained PBDEs. An association between mental and physical development in children exposed to PBDEs has been suggested [71].

What Does All This Mean?

The environmental accumulation of PCBs and PBDEs represent a classic illustration of how industrial chemicals can spread from the site of their usage into air and water, finally ending up in our food and drinking water. Even without the benefit of hindsight it was not totally unexpected because, after all, the antimalarial DDT, which bears some chemical resemblance to the PCBs, did the same thing and, even now, is being detected both in food and water. The assumption that a combination of heat, light, microorganisms, water and other environmental factors would quickly degrade both DDT and the PCBs thereby limiting their build up in the environment has clearly been shown to be just plain wrong. Once again it is difficult to understand how it could not be foreseen that substances like the PCBs, used industrially because of their great chemical stability, would not show these same properties when released into the environment and therefore would degrade only very slowly. It demonstrates, yet again, that despite the accumulating evidence that a particular pollutant may be harmful to the environment and also to humans, governments apparently dismiss much of the early evidence as inconclusive and only act when they are forced to. The obvious conclusion is that once government permission is given for a particular industrial use for a chemical, and there are genuine concerns expressed that it may be harmful, the onus of proof switches to the community who have to show that it is causing harm rather than to the polluter to demonstrate that it is safe. Unfortunately, the level of proof required to satisfy government agencies is considerable, requiring large and expensive studies carried out over many years. During this period of time which, as is amply demonstrated by the length of time taken to ban PCBs, may run into decades, the community and environment continue to be exposed. It is clear that much more stringent regulatory mechanisms are required, and these need to take into account the likely environmental stability of a pollutant and whether it can enter our food chain. There also has to be faster response of governments to scientific evidence that pollutants are likely to be harmful. It is intolerable that the community has to wait for years before action is taken to address its concerns that about safety of a particular industrial waste product. It is also intolerable that the onus of proof regarding the safety of a product rests with the community.

References

1. Hutzinger O et al (1985) Formation of polychlorinated dibenzofurans and dioxins during combustion, electrical equipment fires and PCB incineration. Environ Health Perspect 60:3–9
2. Malisch R, Kotz A (2014) Dioxins and PCBs in feed and food-review from the European perspective. Sci Total Environ 491–492:2–10
3. Gomara B et al (2005) Levels and trends of polychlorinated dibenzo-p-dioxins/furans (PCDD/Fs) and dioxin-like polychlorinated biphenyls (PCBs) in Spanish commercial fish and shellfish products, 1995–2003. Agric Food Chem 53:8406–8413
4. Barone G et al (2014) PCBs and PCDD/PCDFs in fishery products: occurrence, congener profiles, and compliance with European legislation. Food Chem Toxicol 74:200–2005
5. Stefanelli P (2004) Organochlorine compounds in tissues of swordfish (Xiphias gladius) from Mediterranean Sea and Azores islands. Mar Pollut Bull 49:938–950
6. Hamilton MC et al (2005) Lipid composition and contaminants in farmed and wild salmon. Environ Sci Technol 39:8622–8629
7. Hsu JF et al (2005) Congener profiles of PCBs and PCDD/Fs in Yucheng victims fifteen years after exposure to toxic rice-bran oils and their implications for epidemiologic studies. Chemosphere
8. Bates MN et al (2004) Persistent organochlorines in the serum of the non-occupationally exposed New Zealand populations. Chemosphere 54:1431–1443
9. Suzuki G et al (2005) Distribution of PCDDs/PCDFs and Co-PCBs in human maternal blood, cord blood, placenta, milk, and adipose tissue: dioxins showing high toxic equivalency factor accumulate in the placenta. Biosci Biotechnol Biochem 69:1836–1847
10. Fangstrom B et al (2005). A retrospective study of PBDEs and PCBs in human milk from the Faroe Islands. Environ Health 4, 12

11. Deutch B et al (2004) Dietary composition in Greenland 2000, plasma fatty acids, and persisent organic pollutants. Sci Total Environ 331:177–188
12. Denham   M   et   al   (2005)   Relationship   of   lead,   mercury,   mirex, dichlorodiphenyldichloroethylene, hexachlorobenzene, and polychlorinated biphenyls to timing of menarche among Akwesasne Mohawk girls. Pediatrics 115:127–134
13. Quinsey PM et al (1995) Persistence of organochlorines in breast milk of women in Victoria, Australia. Food Chem Toxicol 33:49–56
14. Willhelm M et al (2008) The Duisberg birth cohort study: influence of perinatal exposure to PCDD/Fs and dioxin-like PCBs on thyroid hormone status in newborns and neurodevelopment of infants until the age of 24 months. Mutat Res 659:83–92
15. Cohn BA et al (2012) Exposure to polychlorinated biphenyl (PCB) congeners measured shortly after giving birth and subsequent risk of maternal breast cancer before age 50. Breast Cancer Res Treat 136:267–275. http://www.ncbi.nlm.nih.gov/pubmed/23053646
16. Freeman MD, Kohles SS (2012) Plasma level of polychlorinated biphenyls, non-Hodgkin lymphoma, and causation. J Environ Public Health Epub 2012 April 3. http://www.ncbi.nlm. nih.gov/pubmed/22577404
17. Steenland K et al (2006) Polychlorinated biphenyls and neurodegenerative disease mortality in an occupational cohort. Epidemiology 17:8–13. http://www.ncbi.nlm.nih.gov/pubmed/ 16357589
18. Guo YL et al (2004) Yucheng:health effects of prenatal exposure to polychlorinated biphenyls and dibenzofurans. Int Arch Occup Environ Health 77:153–158
19. Wang SL et al (2003) Neonatal and childhood teeth in relation to perinatal exposure to polychlorinated biphenyls and dibenzofurans: observations of the Yucheng children in Taiwan. Environ Res 93:131–137
20. Dallinga JW et al (2002) Decreased human semen quality and organochlorine compounds in blood. Hum Reprod 17:1973–1079
21. Takser L et al (2005) Thyroid hormones in pregnancy in relation to environmental exposure to organochlorine compounds and mercury. Environ Health Perspect 113:1039–1045
22. Hertz-Picciotto et al (2005) In utero polychlorinated biphenyl exposures in relation to fetal and early childhood growth. Epidemiology 16:648–656
23. Jacobson JL, Jacobson SW (2003) Prenatal exposure to polychlorinated biphenyls and attention at school age. J Pediatr 143:780–788
24. Hardell   L   et   al   (2004)   Increased   concentrations   of   polychlorinated   biphenyls, hexachlorobenzene, and chlordanes in mothers of men with testicular cancer. Environ Health Perspect 111:930–934
25. Viel JF et al (2000) Soft tissue sarcoma and non-Hodgkin's lymphoma clusters around a municipal solid waste incinerator with high dioxin emission levels. Am J Epidemiol 152:13–19
26. Biggeri A, Catelan D (2005) Mortality for non-Hodgkin lymphoma and soft-tissue sarcoma in the surrounding area of an urban waste incinerator. Campi Bisenzio (Tuscany, Italy) 1981–2001. Epidemiol Prev 29:156–159
27. National Toxicol Program Tech Rep Ser 521, 4-232 (2006) NTP technical report on the toxicology and carcinogenesis studies of 2, 3,7,8-tetrachlorobenzo-p-dioxin (TCDD) (CAS No. 1746-01-6) in female Harlan Sprague Dawley rats (Gavage Studies)
28. Jin MH et al (2008) In utero exposure to 2,3,7,8-Tetrachlorodibenzo-p-Dioxin affects the development of reproductive system in mouse. Yonsei Med J 49:843–850
29. Leijs MM et al (2008) Delayed initiation of breast development in girls with higher prenatal dioxin. Chemosphere 73:999–1004
30. Jones-Otazo HA et al (2005) Is house dust the missing exposure pathway for PBDEs? An analysis of urban fate and human exposure to PBDEs. Environ Sci Technol 39:5121–5130
31. Schecter A et al (2004) Polybrominated diphenyl ethers contamination of United States food. Environ Sci Technol 38:5306–5311
32. Johnson-Restrepo B et al (2005) Polybrominated diphenyl ethers and polychlorinated biphenyls in a marine food web of coastal Florida. Environ Sci Technol 39:8243–8250

33. Huwe JK, Larsen GL (2005) Polychlorinated dioxins, furans, and biphenyls, and polybrominated diphenyl ethers in a US meat market basket and estimates of dietary intake. Environ Sci Technol 39:5606
34. Naert C et al (2006) Occurrence of polychlorinated biphenyls and polybrominated diphenyl ethers in Belgian human adipose tissue samples. Arch Environ Contam Toxicol
35. D'Have H et al (2006) Hair as an indicator of endogenous tissue levels of brominated flame retardants in mammals. Environ Sci Technol 39:6016–6020
36. Costa LG, Giordano G (2007) Developmental neurotoxicity of polybrominated diphenyl ethers (PBDE) flame retardants. Neurotoxicity 28:1047–1067
37. Talsness CE et al (2008) In vitro and lactational exposures to low doses of polybrominated diphenyl ether- 47 alter the reproductive system and thyroid gland of female rat offspring. Environ Health Perspect 116:308–314
38. Raldua D et al (2008) First evidence of polybrominated diphenyl ether (flame retardants) effects in feral barbell from the Ebro River basin (NE, Spain). Chemosphere 73:56–64. http://www.ncbi.nlm.nih.gov/pubmed/18597816
39. Fries GF (1985) The PBB episode in Michigan: an overall appraisal. Crit Rev Toxicol 16:105–156
40. Yuan J et al (2008) Elevated serum polybrominated diphenyl ethers and thyroid stimulating hormone associated with lymphocytic micronuclei in Chinese workers from an E-waste dismantling site. Environ Sci Technol 42:2195–2200
41. www.food.gov.uk. Dioxins and dioxin-like PCBs in the UK diet: 2001 total diet survey. July 2003
42. Rudge S et al (2008) Serum dioxin levels in Sydney Harbour commercial fishers and family members. Chemosphere 73:1692–1698
43. Martin L, Klaassen CD (2010) Differential effects of polychlorinated biphenyl congeners on serum thyroid hormone levels in rats. Toxicol Sci 117:36–44
44. Zhang J et al (2010) Elevated body burdens of PBDEs, dioxins, and PCBs on thyroid hormone homeostasis at an electronic waste recycling site in China. Environ Sci Technol 44:3956–3962
45. Goodman JE et al (2010) Weight of evidence analysis of human exposures to dioxins and dioxin-like compounds and associations with thyroid hormone levels during early development. Regul Toxicol Pharmacol 58:79–99. http://www.ncbi.nlm.nih.gov/pubmed/20416351
46. Meeker JD, Hauser R (2010) Exposure to polychlorinated biphenyls (PCBs) and male reproduction. Syst Biol Reprod Med 56:122–131. http://www.ncbi.nlm.nih.gov/pubmed/20377311
47. Cao Y et al (2008) Environmental exposure to dioxins and polychlorinated biphenyls reduce levels of gonadal hormones in newborns: results from the Duisberg cohort study. Int J Hyg Environ Health 211:30–39
48. Golden R, Kimbrough R (2008) Weight of evidence evaluation of potential human cancer risks from exposure to polychlorinated biphenyls: and update based on studies published since 2003. Crit Rev Toxicol 39:299–231
49. Gallagher RP et al (2010) "Plasma levels of polychlorinated biphenyls and risk of cutaneous malignant melanoma". Int J Cancer 9,
50. Hatcher-Martin JM et al (2012) Association between polychlorinated biphenyls and Parkinson's disease neuropathology. Neurotoxicology 33:1298–1304. http://www.ncbi.nlm.nih.gov/pubmed/22906799
51. Golden R, Kimbrough R (2009) Weight of evidence evaluation of potential human cancer risks from exposure to polychlorinated biphenyls: an update based on studies published since 2003. Crit Rev Toxicol 39:299–331
52. Schettgen T et al (2012) Plasma polychlorinated biphenyls (PCBs) levels of workers in a transformer recycling company, their family members, and employees of surrounding companies. J Toxicol Environ Health A 75:414–422
53. Nakatani T et al (2011) A survey of dietary intake of polychlorinated dibenzo-p-dioxins, polychlorinated dibenzofurans, and dioxin-like coplanar polychlorinated biphenyls from food during 2000-2002 in Osaka City, Japan. Arch Environ Contam Toxicol 60:543–545

54. Langer et al (2012) Blood testosterone in middle aged males heavily exposed to endocrine disruptors is decreasing more with HCB and p'p'-DDE related to BMI and lipids, but not with $\sum$15PCBs. Endocrin Regul 46:51–59. http://www.ncbi.nlm.nih.gov/pubmed/22540852

55. El Maidi N et al (2012) Relationship between prenatal exposure to polychlorinated biphenyls and birth weight: as systematic analysis of published epidemiological studies through standardization of biomonitoring data. Regul Toxicol Pharmacol 64:161–176. http://www.ncbi.nlm.nih.gov/pubmed/22735367

56. Park HY et al (2010) Neurodevelopmental toxicity of prenatal polychlorinated biphenyls (PCBs) by chemical structure: a birth cohort study. Environ Health 9:51. http://www.ncbi.nlm.nih.gov/pubmed/20731829

57. Lynch CD et al (2012) The effect of prenatal and postnatal exposure to polychlorinated biphenyls and child neurodevelopment at age twenty four months. Reprod Toxicol 34:451–456. http://www.ncbi.nlm.nih.gov/pubmed/22569275

58. Toxicology and carcinogenesis studies of 2, 3', 4, 4' 5-pentachlorobiphenyl (PCB118) in female harlan Sprague-Dawley rats (gavage studies). Natl Toxicol Program Tech Rep Ser 2010 (559):1–174. http://www.ncbi.nlm.nih.gov/pubmed/21383778

59. Dickerson SM et al (2011) Endocrine disruption of brain sexual differentiation by developmental PCB exposure. Endocrinology 152:581–594. http://www.ncbi.nlm.nih.gov/pubmed/21190954

60. Elabbas LE et al (2011) In utero and lactational exposure to Aroclor 1254 affects bone geometry, mineral density and biomechanical properties of rat offspring. Toxicol Lett 207:82–88. http://www.ncbi.nlm.nih.gov/pubmed/21856390

61. Hirai T et al (2012) Distribution of polybrominated diphenyl ethers in Japanese autopsy tissue and body fluid samples. Environ Sci Pollut Res Int 19:3538–3546. http://www.ncbi.nlm.nih.gov/pubmed/22544599

62. Zota AR et al (2011) Polybrominated diphenyl ethers, hydroxylated polybrominated diphenyl ethers, and measures of thyroid function in second trimester pregnant women in California. Environ Sci Technol 45:7896–7905. http://www.ncbi.nlm.nih.gov/pubmed/21830753

63. Eggesbo M et al (2011) Associations between brominated flame retardants in human milk and thyroid stimulating hormone (TSH) in neonates. Environ Res 111:737–743. http://www.ncbi.nlm.nih.gov/pubmed/21601188

64. Leijs MM et al (2012) Thyroid hormone metabolism and environmental chemical exposure. Environ Health 11(Suppl 1):S10. http://www.ncbi.nlm.nih.gov/pubmed/22759492

65. Beksei JG et al (1987) Immunotoxicology: environmental contamination by polybrominated biphenyls and immune dysfunction among residents of the state of Michigan. Cancer Detect Prev Suppl 1:29–37. http://www.ncbi.nlm.nih.gov/pubmed/2826002

66. Ampleman MD et al (2015) Inhalation and dietary exposure to PCBs in urban and rural cohorts via congener-specific measurements. Environ Sci Technol 49:1156–1164

67. Cok I et al (2012) Analysis of human milk to assess exposure to PAHs, PCBs, and organochlorine pesticides in the vicinity of the Mediterannean city Mersin, Turkey. Environ Int 40:63–69

68. Faroon O, Ruiz P (2015) Polychlorinated biphenyls: new evidence from the last decade. Toxicol Ind Health

69. Fromme H et al (2015) Brominated flame retardants – exposure and risk assessment for the general population. Int J Hyg Environ Health

70. Linares V et al (2015) Human exposure to PBDE and critical evaluation of health hazards. Arch Toxicol 89:335–356

71. US EPA Topical wastes generated by sectors. https://www.epa.gov/hwgenerators/typical-wastes-generated-industry-sectors

Chapter 6
Air Pollutants

While the air we breathe consists of mainly nitrogen, oxygen, carbon dioxide and water vapour, it is no surprise to most people, firstly that there is a lot of other material present and, secondly, that there is considerable variability in the levels of these other materials. For example, the air near a beach, and particularly if there is a heavy surf, contains significant amounts of salt, as well as other substances present in sea water. The well known rusting of automobiles in coastal areas is confirmation of the higher levels of salt in the atmosphere. The smell of a forest, particularly on a warm summer day, or a stream gently flowing through the countryside, or the earth after rain, are all natural smells which we find soothing. Not so soothing are the smells associated with forest fires, volcanic eruptions, or dust storms. Then there are the smells that are generated from human activity. Some, such as the smell of baking bread or a newly mowed lawn are pleasant experiences, but others, such as the smells that relate to mining, agriculture or building activities are not so pleasant. All of these smells, whether they are derived from natural processes or from human activity, are indicative of the presence of volatile or particulate (very tiny solid particles) matter, or a combination of both.

What Are the Main Pollutants in the Air We Breathe?

Air pollution is made up of volatile (ie gaseous material) and particulate matter. The gaseous material includes gases such as nitrogen dioxide (formed from the combustion of fossil fuels or wood), ozone (formed from the interaction of oxygen in the air with nitrogen dioxide and hydrocarbons from fossil fuels), sulphur dioxide (formed from the burning of fossil fuels, smelting, and paper production), carbon monoxide (formed by the burning of fossil fuels), methane (formed by the decomposition of plant matter and produced by livestock)., and hydrocarbons (from motor vehicles). Of course, these are just the main gaseous pollutants, but there are many others that are released as a consequence of industrial activities.

© The Author(s), under exclusive license to Springer Nature Switzerland AG 2021
A. Poulos, *The Secret Life of Chemicals*,
https://doi.org/10.1007/978-3-030-80338-4_6

The particulate matter (or PM) that is a component of air pollution is just as complex. While some is released into the atmosphere as a consequence of natural processes e.g. volcanoes, forest fires, pollen, and moulds, it is the particulate matter that is a by product of human activity, mostly the burning of fossil fuels and other industrial activities, that is of particular concern. These particular PMs, vary considerably in size. For simplification purposes, the PMs have been thought to behave like spheres with varying diameters and masses. The PMs that have been most studied are those with a size of from 1 to 10 μm, where a micrometer is one millionth of a metre, or considerably smaller than the head of a pin. Because of their small size, they are suspended in the air we breathe and hence can find their way into our lungs. Chemical analysis indicates that the particles are made up of materials derived from natural sources such as soil, sand, salt, rocks and dusts, together with materials that originate from industrial activity (dust from cement, sand, concrete, wood dust) and the use of motor vehicles. The latter gives rise to small amounts of metals, and fragments from tyres and brake linings. The PMs can, in turn, bind volatile substances, for example gases formed from the burning of oil, natural gas, or coal, or indeed anything else that is generated by industrial activity and which is capable of interacting chemically with the different particles that are floating in the air. The net result of this is the generation of PMs containing a cocktail of chemical substances. The chemical load carried by the PMs may then be deposited in various parts of our bodies, but particularly our lungs and sinuses.

There is, as well, another type of PM resulting from the chemical reaction of some of the volatile pollutants, for example sulphur dioxide or nitrogen oxides, with each other to form what are referred to as "secondary aerosols". These may be take the form of either liquid droplets or solid particles suspended in the air. The levels of these secondary aerosols varies greatly [32, 33].

The actual amounts of PM in air are highly variable and depend on a variety of factors. For example, the amounts of PMs in heavily industrialized urban air are much higher than in wilderness or country areas [31]. The amounts can also vary according to the time of year and even geographical location [32].

Household Air Pollution

Not all air pollution is associated with industrial activities or motor vehicle exhausts because significant amounts of pollutants are also produced, or introduced, in the home. For example, it has been estimated that almost half the world's population burns biomass for lighting, heating or cooking [23]. Biomass is the term used to describe material derived from living organism and includes wood, coal, dung, crop residues or oil. Even with efficient ventilation systems, small amounts of pollutants are released into the air. If ventilation is poor, as it may be particularly in developing countries, then levels of pollutants can build up to fairly high levels [24]. Analysis of indoor air has also demonstrated the presence of microplastic particles released from clothes containing plastics (e.g. polyester) [43].

Another source of indoor air pollution includes the so-called VOCs or volatile organic components. These are volatile substances that are released from the myriad items found in the home including paints, cleaning agents, cosmetics, furnishings, floor and wall coverings, plastics, and wood. Literally dozens of different chemicals have been shown to be present in the air we breathe in our homes. Perhaps the one that is of most concern is formaldehyde, a potential cancer- inducing substance, but others include pinene, limonene, alcohol, aromatics and hydrocarbons [27, 28].

What Sorts of Chemicals Occur in PMs?

The chemicals that comprise the PMs have been well studied over the last few years. Analysis involves the trapping of the PMs on some sort of filter and then examination of the material that is trapped on the particles. There is an extraordinary mixture of different substances in PMs. They can broadly be divided into two categories – the solid particles and the bound material. The solids were mentioned above and may include traces of salt, metals such as copper, iron, aluminium, zinc, manganese, chromium etc. which are normal components of the earth's crust, or from human activities. The bound material varies according to where samples are taken and may contain substances derived from the burning of fossil fuels, in particular a group of chemicals termed "polycyclic aryl hydrocarbons or PAH, and related substances that also contain the element nitrogen (nitrogen occurs naturally in air) (termed nitro-PAH). Even pesticides, polychlorinated biphenyls (PCBs), and the flame retardants (PBDEs) have been found in PMs [1–4].

It is worth emphasizing that, firstly, PAHs, pesticides, PCBs and PBDEs refer to a class of chemicals (much like the terms "carbohydrate", fat, or protein refers to classes of a particular chemical) and analysis has confirmed that that there are mixtures of each class present in the air. Secondly, these are just some of the many different chemical substances that are present in PMs. In a very real sense, we are all exposed to a true cocktail of pollutants, the composition varying quite dramatically according to what is happening at or near the site of sampling [30].

Are the PMs and Volatile Organic Components Taken up into Our Bodies?

By far the greatest source of PMs is in the air we breathe but it does not necessarily follow that PM components only enter via our lungs. Certainly there is a wealth of information on the deposition of PMs in the lungs of animals such as mice and rats, although the corresponding data for humans is quite limited. In an interesting study carried out on children in Port Pirie which, as mentioned earlier in this book, are exposed to lead through the release of lead mainly into the atmosphere, it was concluded that most of the lead in the children's blood had entered their bodies through their mouths and not the lungs [5]. If that is the case, and given that most of the lead in the soil in towns close to mining activities is believed to have been released into the atmosphere in a particulate form before being deposited, it would appear that little of the atmospheric lead is taken up from the lungs into the body [6].

However, this does not appear to apply to other particulate matter, including the PMs in the air with their diverse chemical cargoes because they have been detected in human lungs [34]. It has also been shown that some of the chemicals bound to PMs are excreted in the urine and for this to happen they must have been taken into the body, into the blood, then the kidneys, and from there into the urine. Studies carried out with animals have demonstrated that deposit in our lungs and it is likely that some of the material bound to PMs will find its way into the bloodstream and thence to throughout the body. In addition, the PMs can also eventually settle as dust on the soil or on our crops, or become entrapped in rain droplets and deposit everywhere, including our rivers and reservoirs, as rain. While high concentrations of PMs are mostly found in urban areas, they can be carried great distances by the wind and eventually settle or wash down with rain many miles from where they were originally formed. PMs can also carried by the wind miles from where they are produced and, over time, will eventually deposit somewhere – on land, sea, rivers, or dams. From wherever they land, it is not that difficult to see how they can get into our food or water and then into our bodies – much like the lead from smelting.

There is no doubt that volatile pollutants, such as for example sulphur dioxide, nitrogen dioxide, and many others that are not necessarily bound to PMs, are also taken up into the body. Animal experiments have confirmed that they are absorbed because chemical changes in many different organs have been detected [36, 37]. Once again, the lungs are probably the main entry point but there are suggestions, based on animal studies, that some of these can enter the body through the mucosa in the nose [35].

What Impact Do PMs and Other Pollutants in the Air Have on Our Health?

Gas Pollutants

There is overwhelming evidence for the detrimental effects of some of the gaseous components of air pollution on human health, particularly at high doses. For example, sulphur dioxide, known to be one of the main contributors to acid rain, combines with water to form an acid which is harmful to the lung. Asthmatics in particular are sensitive to sulphur dioxide [29]. In small amounts, for example in an occupational setting, exposure may even increase the incidence of asthma [7]. Sulphur dioxide is also added to food as a preservative and, in this form, can induce asthmatic responses in asthmatics [30].

Carbon monoxide is a well known poisonous gas which combines with the blood pigment haemoglobin to form a chemically modified form of the pigment, carboxyhaemoglobin, which is unable to carry oxygen. If exposure is prolonged, death results. Ozone causes irritation to the eyes and mucus membranes, and is harmful to the lungs. Because of its very great capacity to react chemically with other substances, it has the potential to alter the chemical structure of other gases, or the constituents of PMs, that may be present in the air. It is being used increasingly to sterilize water, particularly bottled water. It is known that many different substances are produced from the organic materials present in drinking water and some of these have been identified. There is some debate as to whether some of these are harmful. However, the nature of the chemical structures produced by the interaction of ozone with other components in air, and the potential effects on our health remains unknown.

Another gas formed by the combustion of fossil fuels is nitrogen dioxide and its levels can rise significantly in urban areas. There is good evidence that, even in relatively small amounts, it can increase wheezing, shortness of breath, and chest tightness especially in asthmatics and there are even suggestions that it may affect the function of the heart [8, 9].

The gaseous hydrocarbons, including the PAH mentioned above, are byproducts of the combustion of petrol used by motor vehicles, particularly diesel. They are a large class of chemical substances, some of which have been detected in urine indicating that they are taken up from the air. Both the US EPA and the European Union consider that, as a group, the PAH are amongst the most potent cancer producing substances known [38].

PMs

There is ample evidence for the effects of exposure of lung tissue, both in humans and animals, to PMs. It is known that the deposition of PMs in human lungs can increase inflammation. Inflammation is the body's protective response to an injury or trauma. Our immune systems, which play a key role in this process, will destroy, dilute or wall off the injury and are also involved in any subsequent healing. Bacterial or fungal infections are examples of agents that can cause injury and, at the same time, elicit an inflammatory response. On a more simple level, a splinter in the finger generates an inflammatory response. The body's immune cells move to the site of the splinter, special chemicals are released and these in turn increase the capacity of the body to kill any bacteria introduced with the splinter as well as attracting other cells to the site. The permeability or leakiness of blood vessels in the area is greatly increased (this explains the redness), permitting the movement of more white cells and healing substances from the blood. Eventually, the injury is contained, any damage is repaired and, by a mechanism we do not fully understand, the inflammatory process is turned off and everything reverts back to normal. However, if the inflammatory response is not turned off then those same processes which contain and heal an infection, can produce harm. Chronic inflammation i.e. an ongoing inflammatory process is now believed to contribute to many of the degenerative diseases, for example heart disease, arthritis, even cancer, which affect us, particularly as we get older. It follows then that anything that can induce chronic inflammation has the potential to harm us.

In relation to the PMs which are taken up into the lungs, there is considerable evidence that, in animals at least, exposure causes inflammation in the lung and effects on the heart [10, 25]. While it is more difficult to measure inflammatory changes in the lungs of humans, there have been some studies which have confirmed that PMs cause inflammation. In a report published in 2004, PMs from two different areas of Germany were injected into the lungs of healthy volunteers and lung fluid removed 24 h later [11]. Analysis of the fluid demonstrated that two well known markers of inflammation were present in higher concentrations in the lungs of the volunteers that had been injected with PMs from an area of Germany (Hettstedt) known to have a higher than normal incidence of allergic asthma. It was speculated that the PMs possibly contributed to the allergic asthma, perhaps by inducing chronic inflammation in the lungs. Indeed, there is evidence that air pollution does contribute to the development of asthma although more recent research has suggested that any association may be limited to more vulnerable groups in the community, such as the elderly [18, 41, 42]. However, we know that exposure to PM, as well gaseous pollutants such as ozone and nitrogen dioxide can reduce the lung peak respiratory flow (PRF) [15]. The PRF is the measurement of the rate at which air is expelled from the lungs and is believed to correlate with inflammatory changes in the organ [15]. Other researchers have demonstrated that PM exposure increases the production of nitric oxide, a gas formed by the lungs and associated with inflammation [16]. What is particularly alarming is that chemical indicators of inflammation after exposure to air pollution are not only found in the lungs but also in the blood [26].

Apart from its effects on the function of the lung, air pollution is believed to contribute to diseases of the blood vessels and heart (cardiovascular disease) [12, 17, 39, 40]. This includes heart attacks, heart failure, arrhythmias (abnormalies in heart rhythm), strokes, and even deep vein thrombosis. It has been suggested that, world wide, air pollution, and PMs in particular, may contribute to the death of as many as 800,000 people per year and that some people, notably asthmatics and the elderly are particularly susceptible [22]. There is considerable debate as to how PMs do this. One theory is that PMs enter blood circulation and induce inflammation followed by subsequent damage to the blood vessel walls (believed to contribute to hardening or sclerosis of the arteries), another is that the particles can alter the autonomic nervous system activity (that part of the nervous system that regulates the heart). Certainly some of the components of diesel exhaust can increase the formation of thrombus, the material that deposits in blood vessels and leads to blockage followed by either strokes or heart attack [13].

There are also indications that air pollution may contribute to the development of a variety of cancers, including those affecting the lung, bladder, cervix and brain, and leukemias [14, 19–21]. However, much of the evidence for this is based on epidemiological studies which look at the incidence of cancer in large populations exposed to varying amounts of air pollutants. If the incidence of cancer increases with greater exposure, and providing there are no other factors that can explain the cancer, then air pollution is considered to have contributed to the risk. The problem with this type of study is that it is not definitive but only suggests a possible relationship between pollution and disease. There are a great number of variables – diet, smoking, alcohol consumption, and workplace exposure to chemicals are just a few. And, while it is possible to correct for these, and scientists carrying out the studies do this routinely, there are many other factors that may or may not contribute to the fact that one group has a higher incidence of disease than another. Moreover, when groups with differing pollution exposures are compared, it does not take into account that individuals within each group are exposed to a great number of different substances that are present in the air they breathe. It is exceedingly difficult to quantify all of these so, when scientists put individuals into groups according to the degree of air pollution they tend to select one of the major pollutants, such as nitrogen dioxide, and assume that all people in this group have a higher exposure to all pollutants in the air. It may be true for nitrogen dioxide but clearly does not apply to whatever else may be in the air.

What Does All This Mean?

There is no question that we are all exposed, to some degree or other, to air pollutants, in our homes and outside in the "fresh air". The pollutants exist in either a gaseous form, or as solid particles which may or may not contain bound materials. The latter, the so-called PMs which vary greatly in size, have generated the most concern because it has been shown that they carry a mixture of potentially harmful substances. Some of the pollutants are produced naturally but many are derived from industrial activities, motor vehicles, household heating, and power generation to name just a few. People who live in urban areas are the most exposed although pollutants can travel many miles in the air and end up even in rural areas where there is no industrial activity. In a way it is like the release of pollutants such as PCBs or mercury into the sea at a particular site, and the detection of these contaminants in animals such as fish and whales perhaps thousands of miles away from the site of contamination.

There is also no question that the various pollutants are taken up into our lungs where they may induce inflammation of the delicate lung tissue and increase the risk of asthmatic attacks in asthmatics. Chemical markers of this increased inflammation are not only found in the lung but are also present in the blood. It is generally believed that air pollution can increase the risk of cardiovascular diseases, such as heart attacks and strokes, and there are also indications that the risk of cancers such as bladder and lung cancer, as well as leukemia, are increased on continuing exposure to air pollution. Governments are aware of the threats posed by global pollution – the global warming debate is not just about carbon dioxide – and have introduced measures to limit pollution e.g. alternative energy sources, stricter industrial emission controls, but really effective action is difficult because of the rapid industrialization of countries like India and China, the long lead times required to introduce newer less polluting technologies, and the perception that being green is synonymous with job losses. There is also the view, never expressed overtly, that air pollution is the price that we all have to pay if we want jobs and a better lifestyle. Unfortunately, if the latter view continues to prevail, some cataclysmic event will be required before governments are forced to confront the threat of air pollution.

References

1. Wang X et al (2008) A wintertime study of polycyclic aromatic hydrocarbons in PM(2.5) and PM(2.5-10) in Beijing: assessment of energy structure conversion. J Hazard Mater 157:47–56
2. Wang X et al (2008) Organochlorine pesticides in particulate matter of Beijing, China. J Hazard Mater 155:350–357
3. Cheng JP et al (2007) Polychlorinated biphenyls (PCBs) in PM10 surrounding a chemical industrial zone in Shanghai, China. Bull Environ Contam Toxicol 79:448–453
4. Deng WJ et al (2007) Distribution of PBDEs in air particles from an electronic waste recycling site compared with Guangzhou and Hong Kong, South China. Environ Int 33:1063–1106
5. Kranz BD et al (2004) The behaviour and routes of lead exposure in pregrasping infants. J Expo Anal Environ Epidemiol 14:300–331
6. Tembo BD et al (2006) Distribution of copper, lead, cadmium and zinc concentrations in soils around Kabwe town in Zambia. Chemosphere 63:497–501
7. Andersson E et al (2006) Incidence of asthma among workers exposed to sulphur dioxide and other irritant gases. Eur Respir J 27:720–725
8. Belanger K et al (2006) Association of indoor nitrogen dioxide exposure with respiratory symptoms in children with asthma. Am J Respir Crit Care Med 173:297–303
9. Felber Dietrich D et al (2008) Differences in heart variability associated with long term exposure to NO2. Environ Health Perspect 116:1357–1361
10. Happo MS et al (2007) Dose and time dependency of inflammatory responses in the mouse lung to urban air coarse, fine, and ultrafine particles from six European cities. Inhal Toxicol 19:227–246
11. Schaumann F et al (2004) Metal-rich ambient particles (particulate matter 2.5) cause airway inflammation in health subjects. Am J Respir Crit Care Med 170:898–903
12. Brook RD (2008) Cardiovascular effects of air pollution. Clin Sci 115:175–187
13. Lucking AJ et al (2008) Diesel exhaust inhalation increases thrombus formation in man. Eur Heart J 29:3043–3051
14. Liu CC et al (2009) Ambient exposure to criteria air pollutants and risk of death from bladder cancer in Taiwan. Inhal Toxicol 21:48–54. http://www.ncbi.nlm.nih.gov/pubmed/18923949
15. Correia-Deur JE et al (2012) Variations in peak expiratory flow measurements associated to air pollution and allergic sensitization in children in Sao Paulo, Brazil. Am J Ind Med 55:1087–1098. http://www.ncbi.nlm.nih.gov/pubmed/22544523
16. Strak M et al (2012) Respiratory health effects of airborne particulate matter: the role of particles size, composition and oxidative potential: the RAPTES project. Environ Health Perspect 120:1183–1189. http://www.ncbi.nlm.nih.gov/pubmed/22552951
17. Krishnan RM et al (2012) Vascular responses to long and short-term exposure to fine particular matter: the MESA (multi-ethnic study of arteriosclerosis and air pollution). J Am Coll Cardiol. http://www.ncbi.nlm.nih.gov/pubmed/23103035

18. Portnov BA et al (2012) High prevalence of childhood asthma rjr Israel is linked to air pollution by particulate matter: evidence from GIS analysis and Bayesian model averaging. Int J Environ Health Res 22:249–269. http://www.ncbi.nlm.nih.gov/pubmed/22077820

19. Raaschou-Nielson O et al (2011) Lung cancer incidence and long-term exposure to air traffic pollution from traffic. Environ Health Perspect 119:860–865. http://www.ncbi.nlm.nih.gov/pubmed/21227886

20. Raaschou-Nielsen O et al (2011) Air pollution from traffic and cancer incidence: a Danish cohort study. Environ Health 10:67. http://www.ncbi.nlm.nih.gov/pubmed/21771295

21. Amigou A et al (2011) Road traffic and childhood leukemia: the ESCALE study (SFCE). Environ Health Perspect 119:566–572. http://www.ncbi.nlm.nih.gov/pubmed/21147599

22. Anderson JO et al (2012) Clearing the air: a review of the effects of particulate matter air pollution on human health. J Med Toxicol 8:166–175. http://www.ncbi.nlm.nih.gov/pubmed?term=%22anderson%20jo%22%20AND%20%22clearing%20the%20air%22

23. Diette GB et al (2012) Obstructive lung disease and exposure to burning biomass fuel in the indoor environment. Glob Heart 7:265–270. http://www.ncbi.nlm.nih.gov/pubmed/23139916

24. Dasgupta S et al (2006) Indoor air quality for poor families: new evidence from Bangladesh. Indoor Air 16:426–444. http://www.ncbi.nlm.nih.gov/pubmed/17100664

25. Amatullah H et al (2012) Comparative cardiopulmonary effects of size-fractionated airborne particulate matter. Inhal Toxicol 24:161–171. http://www.ncbi.nlm.nih.gov/pubmed/22356274

26. Dutta A et al (2012) Systemic inflammatory changes and increased oxidative stress in rural Indian women cooking with biomass fuels. Toxicol Appl Pharmacol 261:255–262. http://www.ncbi.nlm.nih.gov/pubmed/22521606

27. Cheng M et al (2016) Factors controlling volatile organic compounds in dwellings in Melbourne, Australia. Indoor Air 26:219–230

28. Mendell MJ (2007) Indoor residential chemical emissions as risk factors for respiratory and allergic effects in children: a review. Indoor Air 17:259–277

29. Reno AL et al (2015) Mechanisms of heightened airway sensitivity and responses to inhaled SO2 in asthmatics. Environ Health Insights 9(Suppl 1):13–25

30. Vally H, Misso NL (2012) Adverse reactions to the sulphite additives. Gastroenterol Hepatol Bed Bench 5:16–23

31. Shi GL et al (2015) Further insights into the composition, source, and toxicity of PAHs in size resolved particulate matter in a megacity in China. Environ Toxicol Chem 34:480–487

32. Rhode RA, Muller RA (2015) Air pollution in China. Mapping of concentration and sources. PLoS 10(8)

33. Guo S et al (2014) Elucidating severe urban haze formation in China. Proc Natl Acad Sci U S A 111:17373–17378

34. Churg A, Brauer M (1997) Human lung parenchyma retains PM2.5. Am J Respir Critic Care Med 155:2109–2111

35. Satou T et al (2013) Organ accumulation in mice after inhalation of single or mixed essential oil compounds. Phytother Res 27:306–311

36. Xie J et al (2007) Protein oxidation and DNA-protein crosslink induced by sulfur dioxide in lungs, livers, and hearts of mice. Inhal Toxicol 19:759–765

37. Han M et al (2013) Nitrogen dioxide inhalation induces genotoxicity in rats. Chemosphere 90:2737–2742

38. Ifegwu OC, Anvakora C (2015) Polycyclic aromatic hydrocarbons: part 1. Exposure. Adv Clin 72:277–304

39. Lu F et al (2015) Systematic review and meta-analysis of the adverse effects of the ambient PM2.5 and PM10 in the Chinese population. Environ Res 136:196–204

40. Lee BJ et al (2014) Air pollution exposure and cardiovascular disease. Toxicol Res 30:71–75
41. Jacquemin B et al (2015) Ambient air pollution and adult incidence in six European cohorts (ESCAPE). Environ Health Perspect 123:613–621
42. Simoni M et al (2015) Adverse effects of outdoor pollution in the elderly. J Thorac Dis 7:34–45
43. Dris R et al (2017) A first overview of textile fibers, including microplastics, in indoor and outdoor environments. Environ Pollut 221:453–458

Chapter 7
Water Everywhere – But Is it Safe to Drink?

What it is that makes up as much as 70% or more of the human body, is found in all living things and is essential for all life forms on the planet, changes from a vapour, to a liquid and then solid at different temperatures, and has a unique ability to dissolve an extraordinary variety of substances? There are no prizes for guessing that the substance is water. It is in the food we eat, in the air we breathe, and the liquids we drink. It is absolutely everywhere.

Water is quantitatively the most important component of the human body. Its proportion gradually decreases from around 88% of total body weight in a 20–25 week foetus, to 69% in a full term baby, and 60% in an adult male [1]. All of our organs are literally made of water. A 70 kg man is made up roughly of about 45 litres of water. The myriad chemical reactions taking place in our various organs take place in an environment of water, and the blood that carries nutrients to various parts of the body as well as waste products is also largely water. Just about everything we eat or drink contains water. On a daily basis, perhaps as much as one litre of water or more is taken in as the water component of the food we eat, and another litre or so water as a drink, either water itself, tea, coffee, alcoholic drinks etc. [2].

The major source of water in most cities around the world is rain although increasingly desalination of sea water is used if rainfall is insufficient. When it rains in country areas the water is taken up by the soil until saturation and then, if the land is higher than the adjoining land, it will run downhill to a lake, river, pond, or even the ocean. Some of the water is diverted to dams which in turn provide the water supply for cities and townships.

As mentioned above, one of the most important properties of water is its capacity to dissolve a great variety of substances- solids, liquids, or even gases. The oceans and seas of the world contain large amounts of dissolved salts, mainly common salt although there are many other substances as well. Rivers, a major source of drinking water for many people, also contain varying amounts of dissolved substances, as well as sediment derived from rocks and soil. Sediment is formed by the physical action of water as it flows over a surface and can include particles of clay, sand, and minerals in addition to tiny remnants of animal and plants. Included in the sediment increasingly are tiny particles of plastics, referred to as microplastics or microfibres formed by the breakdown of plastics or release from textiles [3]. River water also naturally contains soluble organic matter (ie carbon containing substances) which will be discussed in greater detail below.

Ocean and river waters also contain a number of dissolved gases, including oxygen, carbon dioxide, and nitrogen, the main gases present in air. The level of oxygen in particular is critical to sea and river animals because reduction of normal levels can cause disease and death. Even gases such as methane, present close to oil and gas reserves, and also formed by the microbial decomposition of organic matter, and hydrogen sulphide, may be found in water.

Reservoir Water

As water is a supreme solvent it is not surprising that, whatever the source, it is unlikely to be completely free of contaminants. Even water that flows from the watersheds comprised entirely of pristine native forests into reservoirs contains organic matter and other dissolved and particulate materials mentioned above and therefore has to undergo rigorous purification and disinfection processes prior to consumption. Natural contaminants in reservoir water include organic matter mentioned above ie decaying leaves, wood, wildlife excreta, as well as inorganic material ie mainly non carbon containing substances ie fine particles of sand and rocks, as well as mineral salts which have been dissolved by the water as it flows over the land. The main mineral salts include sodium, calcium, iron, magnesium, chromium, manganese, aluminium, and copper salts. Depending on the composition of the soil and any underlying rocks there may also be small amounts of the more toxic metals such as lead, arsenic, and cadmium, and even radioactive elements such as uranium and thorium. As mentioned earlier in this book, arsenic is a common component in ground water from many parts of the world because of its natural presence in the soil from geothermal activities and weathering of arsenic-containing rocks. While the levels of arsenic in natural water are mostly less than 10 millionths of a gram per liter, the permitted World Health Organisation limit, in some parts of the world, for example Pakistan, the levels may be greater than 50 micrograms per litre and, in some cases, levels of greater than 300 millionths of a gram per litre [4].

There are, as well, a number of other natural factors than can affect the composition of water flowing from the different catchments into reservoirs. Weather conditions, such as season of the year, droughts or floods, can change both the level and composition of organic matter as well as the mineral content of the runoff from the watershed. Similarly, bushfires which wreak havoc on the vegetation can increase runoff and, in turn, can lead to greater amounts of soil, ash and dissolved organic matter in the water. The heat from the wildfires can also change the chemical structures of some of the components in the soil and these, in turn, may find their way into reservoirs [5].

In addition to the host of naturally occurring metals and other substances present in leaves and other debris that finds its way into reservoirs, there are many other substances that are byproducts of human activities. These include fertiiisers, animal manures, antibiotics, pharmaceuticals, sunscreen filters, synthetic sweeteners (cyclamate, saccharin, sucralose), industrial chemicals such as the fluorocarbons, perfluorooctane sulphonate (PFOS) and perfluorooctanoic acid (PFOA), personal care products, flame retarding chemicals such as PBDE, plastics components such as bisphenol A, and pesticides have been reported in untreated water from different sources. Traces of chemicals released from myriad human activities are found in river water and sediment even in remote areas [6–11]. Even radioactive elements, such as those released by mining, industry, medical procedures, research can find their way into water [12].

Extremes of weather, such as bushfires and floods, can also lead to the contamination of reservoirs, recreational lakes and rivers with large amounts of certain nutrients, particularly phosphorus which, if not removed, could cause algal growth and the production of toxins. In these situations, a first step in the treatment of water may involve the use of a lanthanum-clay mixture to remove excess phosphorus [13].

Apart from dissolved materials and sediment, river water or the fresh water of reservoirs, can also contain organisms that can cause diseases in humans. Examples include bacteria that can cause typhoid, cholera, and gastroenteritis, and parasites that cause diseases such as amoebic dysentery, giardiasis, and malaria, as well as enteric viruses, that cause diseases such hepatitis, polio, influenza, as well as parasites that can cause diseases such as amoebic dysentery, giardiasis and malaria [14, 15]. There is little evidence that the coronavirus is present in surface or groundwaters and transmitted through contaminated drinking water [16]. Most of these organisms, termed "pathogens", usually come from contamination with human, livestock, and even wildlife excreta. Seasonal changes in the water levels in reservoirs can also influence the levels of cyanobacteria, an organism that produces toxins that can affect the function of the liver, kidney, nervous system, reproductive organs, and brain [17].

Stormwater

In cities, much of the rain falls on impervious surfaces such as roads, railways, car parks, and roofs and from there into drains then rivers. Creeks, and the ocean. However, as it travels it carries a variety of debris, such as cigarette butts, plastics, paper, oil and grease, as well as dissolved materials and microorganisms, that are present on these various surfaces the composition varying according to land use in the area. Despite this, stormwater is being increasingly considered a valuable resource and is being used in industry and agriculture and for irrigating public areas. Before use, trash is removed followed by a filtration process to remove particulate matter. Depending on the intended use further treatment may be required to remove chemical and biological contaminants [18, 19].

Graywater

Graywater is the water that is generated in households from bathing, washing machines, and dishwashers but not from toilets [20, 21]. Depending on its intended use- not normally for drinking- it may require filtration and treatment with disinfection agents. Greywater has been used without any treatment for toilet flushing or watering the garden although, because of the presence of numerous chemical pollutants, such as detergents, pharmaceuticals, health and beauty products, aerosols etc. as well as microorganisms, biological, microfiltration, ultrafiltration, and adsorption processes are mostly used for treatment prior to use. As part of the latter process a range of naturally occurring media, including activated charcoal, peat moss, lime pebbles, and pine bark may be used [21].

Sewerage

A significant proportion of water used per household is present in sewerage. It is defined as all waste water used for bathing, washing, toilet flushing that flows into a septic tank or into drains and from there to a community sewerage system. Sewerage comprises water as well as solid material or sludge. The latter includes faeces and urine, so contains microorganisms such as bacteria, fungi, and even viruses. Indeed, the testing for viruses, such as the SARS-CoV-2 virus, at different sites within the sewerage system is thought will provide a means of determining likely sites of infections within the community [22, 23]. Sewerage also contains a mixture of chemicals including food components for example protein, carbohydrate and fats including their respective breakdown products, food additives, pharmaceuticals, and supplements. It is worth noting that each of these classes may contain mixtures of chemicals. For example pharmaceuticals comprise a great variety of different chemical substances including anti-inflammatories, antibiotics, antiphysychotics, antidiabetics, antihypertensives, and statins [24]. In addition to the release of microplastics from the washing of clothes, water derived from dishwashers, laundry, and bathing, may contain a variety of chemicals. Including surfactants, builders, corrosion inhibitors, fluorescers, bleaches, buffers, polymers, chelants, suds suppressors, fragrances, dyes, perfumers, and preservatives [25]. As there is limited information on detergent packets regarding the chemical composition it is useful to look at the composition of some typical dishwashing detergents for which there are granted patents and shown on the US patent database site. For example, the chemical complexity of a typical dishwashing detergent is shown in US patent number 10, 597, 611 which may include surfactants, solvents, mineral bases, builders, and co-builders [26].

Because of presence of microorganisms and a variety of chemical pollutants, sewerage requires rigorous purification processes before use. It is a complicated process involving primary, secondary and tertiary treatments. The primary treatment involves the use of filters to remove solid material such as cotton buds, rags, paper, and rubbish, treatment with oxygen to assist in the floatation of sand and grit, and sedimentation which generates sludge. Secondary treatment involves bacterial aerobic (in the presence of oxygen) and anaerobic (in the absence oxygen) digestion for the breakdown of organic materials. After any sludge is allowed to settle in large sedimentation tanks, in the final stage (tertiary treatment) the resulting liquid is treated with disinfecting agents such as ozone (a chemically modified form of oxygen), chlorine, and ultraviolet light. Much of the recycled water from sewage plants is not used directly as a source of water directly but indirectly through its discharge into water catchment rivers and from there it undergoes the usual purification and disinfection processes. The sludge and solid material, referred to as biosolids, will be discussed below.

Tank Water

Rainwater flowing onto the roofs of buildings can be used as a source of drinking water particularly in country areas. It flows from the roof into gutters then into tanks. In most instances it is assumed that tank water is chemical free and safe to drink. However, unless the roof and gutters are cleaned regularly, the water in the tank may contain organic matter from leaves and bird droppings, as well as dirt and dust that can accumulate on the roof surface during periods without rain. In addition, particulate matter normally present in the air in cities, and even country areas, and containing a variety of pollutants may be carried by the rain onto roofs and then into tanks. Yet another source of tank contaminants include the components that comprise the roof and gutters, for example the paints, metals or tiles and their breakdown products that build up over periods of time. Finally, a great variety of bacterial microorganisms including E. Coli, Salmonella, Legionella, and Pseudomonas have been detected in tank water [27]. It is clear that what is on the roof can end up in tank water and, ultimately, taken up into our bodies.

Bottled Water

The US FDA classifies bottled water according to whether it is artesian well water, mineral water, or spring water [28]. There are a variety of sources for bottled water including well water, artesian well water, spring water, desalinated water, or mineral water. In addition, some bottled water may be tap water which is treated before bottling by any one of a variety of methods including distillation, reverse osmosis, or filtration. Distillation involves heating water to boiling and collecting and condensing the water vapour. Reverse osmosis involves the use of special membranes to remove dissolved solids [28]. Because of the concerns about contamination with microorganisms tap water is often disinfected with ozone gas prior to bottling. There is increasing evidence that, depending on its source and packaging, bottled water may contain contaminants including microplastics and plasticisers such as bisphenol A with estrogenic activity, phthalates, and PFAS [29–32]. In addition, flavouring enhancers may be added to bottled water and this can increase the acidity of the water which, in turn, may increase the risk of tooth decay [33].

Rainwater

Most people would be surprised to learn that even rain is not entirely free of environmental contaminants [34]. Many of the chemicals associated with industrial and other activities, such as aromatic polycyclic hydrocarbons, polychlorinated biphenyls, and pesticides have been found in rain. It is likely that environmental chemicals are transported from their source in the air and find their way into water droplets in clouds and then into rain. Significant amounts of microplastics can also occur in the atmosphere and, from there, find their way into rain [35].

Desalinated Water

In times of drought, or in countries with limited natural water supplies, water from the ocean, seas, or saltwater lakes are potential sources of water. The problem, of course, is the presence of various salts, mainly common salt, which have to be removed before the water is fit to drink. Salts can be removed from water by distillation which involves heating the water to boiling then cooling the water vapour by condensation. The dissolved salts are mostly not volatile so the condensed water is relatively pure. At the same time, any microorganisms that may be present in the water are destroyed at the high temperature so the distilled water is mostly safe to drink. While this process is effective, it requires large amounts of energy.

Another way water may be desalinated is through reverse osmosis. This involves the use of special membranes which are permeable to water but not to any dissolved salts. Once again, significant amounts of energy are required, this time to drive the water through the membranes. While the method is effective, small amounts of salts may be present in the purified water. The process is effective in removing microorganisms which are mostly larger than the membrane pores sizes although because of the risk of bacterial contamination of the membranes and other components of the equipment used (referred to as "biofouling"), disinfection is mostly required [36]. The types and concentrations of disinfection byproducts depend on several factors, such as the type and amount of disinfectant used, the contact time, the organic and inorganic contents, the temperature, and the pH ie the acidity or alkalinity. There is also an environmental impact on the use of these membranes because they have a limited life and many are discarded in landfills [37]. Whichever method is used, desalination generates large amounts of salinity which have been reported to have effects on some marine organisms [38].

Desalinated water is commonly consumed in Middle Eastern countries and other countries with limited fresh water supplies. There are some concerns because the procedure may result in the removal of certain nutritionally important mineral components of drinking water found in other sources, in particular iodine, magnesium, and calcium [39–41].

Tapwater

It is generally assumed that many of the contaminants present in water are removed by treatment. However, while most are removed, analysis of drinking water in different parts of the world has demonstrated that it is not always complete and traces of some contaminants, albeit at very low concentrations, may still be detected [42, 43]. It is also worth noting that additional contaminants can be introduced during transport from the water treatment site to homes, industry, and other sites. This will be discussed in greater detail later.

Groundwater

Groundwater is water that is located below the ground rather than above the ground such as, for example rivers and lakes. Permeable rocks below the ground containing water are referred to as aquifers. The water in aquifers may arise from the movement of surface water from rain, rivers or streams to the ground below. Aquifers may be recharged using water from different sources [44]. The water in aquifers is accessed in many parts of the world via the construction of wells. The contaminants present in groundwater may include metals such as arsenic released from natural sources such as minerals containing arsenic, cadmium and fluoride, as well as others generated by human activities, for example antibiotics, pharmaceuticals, and pesticides.

In more recent times, hydraulic fracturing of underground rocks to release hydrocarbons eg gas and oil, referred to as fracking, is yet another potential source of contamination of groundwater. The contaminants of groundwater near fracking sites may include gases such as methane and other hydrocarbon gases, and various chemicals that may be added to a frack such as acids, friction reducers, corrosion and scale inhibitors, lubricants, and biocides [45].

Recycled Water

As the population grows, and as in some years there is a reduction in rainfall, there is an increasing need for the reuse or recycling of water. Recycled water is currently being used in a number of areas including agriculture, firefighting, irrigation of public parks, industrial uses such as cooling, washing, and toilet flushing [46]. The main sources of recycled water include sewerage, graywater, and stormwater. As water from these various sources can contain a complex mixture of chemicals, as well as microbiological contaminants (eg bacteria, viruses, protozoa, fungi), substantial purification, followed by disinfection, is required before usage [47]. Recycling of water, in particular stormwater, may also generate large amounts of solid waste which are either recycled or deposited as landfill, while recycling of sewerage generates some waste but it can serve as a source of valuable biosolids which have a number of uses, particularly as a fertiliser.

Biosolids

The solid material produced from sewerage after removal of much of the water, ie the biosolids, has a variety of uses. Including in agriculture as composts and fertilisers, construction materials, and rehabilitation of land such as mining sites. The amounts produced around the world are significant. For example, around 317, 00 tons was produced in Australia in 2017 of which 75% was applied to land. The amounts of biosolids produced in other countries is even higher in the USA, United Kingdom and Japan (17.8, 1.05 and 2.2 million tonnes respectively) [48, 49]. The number of potential environmental contaminants present in sewage sludge and biosolids is large and includes pharmaceuticals, antibiotics, endocrine disrupting substances, PBDEs, and PFOS [49, 50]. Of increasing concern is the presence of microplastics, tiny plastic particles, derived from a variety of plastics such as polypropylene, polyethylene, polyvinylchloride etc. which are present in many different products eg toothpaste, body washes, and personal care products. It has been estimated that biosolids added to soil in agriculture may contain as much as 0.9–2.5% of microplastics [51].

How Is Reservoir Water Made Safe for Drinking?

Given that raw water is often contaminated with sediment and potentially harmful chemicals and organisms that may cause disease, there is clearly a need for purification and disinfection before the water is safe to drink [52]. It is a fairly complex process often involving multiple steps. However, depending on the quality of the water prior treatment may be required before the purification and disinfection steps.

For example, if algae are present in the reservoir water in significant amounts, to reduce the risk of algal growth and their production and release of toxic chemicals, algicides such as copper salts may be added to the water prior to further treatment [52]. Excessive phosphate in water may also be a problem because it too can also lead to excess algal and plant growth and potential release of toxins. To remove excess phosphate, a modified bentonite clay containing the metal lanthanum, and termed Phoslock, may also be added prior to further treatment [52, 53].

The generation of drinking water involves firstly the removal of any particulate matter or sediment because even if the source of water is from protected catchments where there is little agriculture or other human activities, it will contain some particulate matter. It is first held in a holding reservoir where some of the sediment settles out. To remove most of the remaining sediment which may contain smaller particles, flocculating agents, often aluminium salts such as aluminium sulphate, are then added. The object of this treatment is to coalesce the suspended particles into larger particles so that they can more readily be removed by filtration. Depending on the ease or otherwise of flocculation, coagulation aids may be used. There a number of coagulation aids used, including relatively simple substances such as calcium oxide and sodium silicate, but also the more complex polymeric substances, such as polyDADMAC and polyacrylamides which are types of plastic resin. Finer sediment is then removed by passage through beds of sand or charcoal. The latter can also bind chemical contaminants such as pesticides and herbicides. The filtration step may also remove some, if not most, of any remaining pathogens. Other chemicals may be added to water to reduce the acidity or soften the water. For example, hard water contains excess amounts of calcium and magnesium salts and can result in problems with the use of soaps and detergents. It may be corrected by the addition of a relatively harmless salt, sodium carbonate. The last step is the disinfection process which destroys any microorganisms likely to cause disease. This will be discussed in some detail below.

Despite the rigorous purification processes used, small amounts of a variety of different substances remain in the water. A typical analysis of drinking water could include the following components-

Metals/metallic salts – calcium, iron, arsenic, chromium, aluminium, copper, potassium, sodium, zinc, manganese

Non metallics – silica, selenium, boron, bromides, fluoride, selenium

Dissolved organic substances

Dissolved solids

Halomethanes

The dissolved solids, mostly in the form of metallic salts, and non metallics are mostly natural components of soil and rocks over which the water flows from the watershed to the reservoir. The metallic salts are mostly sodium, magnesium, and potassium salts, chemically linked to non metallics, for example oxygen, nitrogen (eg nitrates), sulphur, carbon, bromine, and fluorine. Amounts of metallic salts may reach as much 100 mg per litre. A good example of a metallic salt is ordinary table salt which is a mixture of sodium (a metal) and chlorine (a gas). Small amounts of fluoride are naturally present in water but additional amounts are often added to improve oral health. Halomethanes are products formed by the interaction of the most common disinfection agent used (chlorine) with some of the dissolved organic substances and will be discussed below. It is important to note that the amounts of the various components can vary significantly according to the source of the water ie the reservoir from which the water is obtained. The variation in concentrations in milligrams per litre of some of the components of water taken from different sources for Melbourne water is shown below [54].

Calcium – 3–10
Dissolved solids – 31–95
Nitrate – 0.03–0.45
Sodium– 3.5–15
Total organic carbon – 1.2–3.0
Total trichloroacetic acid – 3–48 (micrograms per litre)
Total dichloroacetic acid – less than 1–21 (micrograms per litre)

The "total organic carbon" refers to any carbon containing substances, either naturally occurring or resulting from human activities. Most of the organic carbon in our drinking water is naturally occurring and contains a great variety of different chemical substances derived from animal, plant, fungal, and bacterial sources. The actual amounts of these different substances can vary significantly. Considerable research has been devoted to determining the chemical structures of the materials comprising what is termed "total organic carbon". Put simply, total organic carbon is made up of substances containing one or more carbon atoms. All living things are made up of a great variety of organic substances. Sugar, starch, and alcohol are examples of organic substances made by living things. While we, and all living organisms, are made up of organic substances, humans have also made literally thousands of different organics which we use for many different purposes. Total organic carbon therefore includes naturally occurring substances that are present in, or made by, living things (including humans), but it may also include substances made by humans for a variety of different purposes eg pesticides, industrial chemicals, pharmaceuticals, detergents etc. Humic and fulvic acids are also a part of the total organic carbon in the organic matter present in soils which eventually finds its way into water. They are highly complex chemical structures formed from microbial action on plant and animal matter contained in soil. The term "total organic carbon" is used to indicate how much of a complex mixture of carbon-containing substances is present in water [55]. Assuming that up to 50% or more of the organic matter that is present in drinking water occurs as carbon, and based on the table shown above, there is up to 6 mg/litre of a mixture of a great number of carbon containing substances in drinking water [56]. Some of these are naturally occurring while others, like the trihalomethanes, have been formed by the disinfection process. As most of us drink between one and two litres of water per day, every day, in some form or other, the total amounts consumed are not insignificant.

The Disinfection Process

As water is Nature's most potent solvent with the capacity to dissolve a variety of chemicals, it is clear that even reservoir water, from the most pristine sources, contains a complex mixture of organic and inorganic chemicals, the composition varying according to a number of factors some of which are mentioned above. Even after the normal purification processes small amounts of contaminants remain in amounts considered to be harmless. Some of these may not react with the chemical agents used to disinfect water. On the other hand, many of the chemical contaminants do so the net result of any disinfection process is the formation of a complex mixture of disinfection products and naturally occurring chemicals. While it is possible to predict some of the likely products, the chemical complexity of the organic components of water from different reservoirs is such that determination of all of the disinfection byproducts is exceedingly difficult. Some of these products may be formed in significant amounts, while others may be present in sub-microgram amounts per litre.

Chlorination

The most commonly used process used for disinfection is referred to as "chlorination" and involves the use of chlorine gas, or sodium or calcium hypochlorite, both of which are sources of chlorine. This process is similar to that used to treat swimming pools. Chlorine itself is a highly reactive, toxic gas. It belongs to a group of highly reactive substances which are termed "halogens". Two other examples of halogens are bromine and fluorine. Chloramine, a chlorinated derivative of ammonia, is also used. Chlorine dissolved in water is a very powerful oxidizing agent ie it is able to change the chemical structure of many substances and it is this property that makes it such a potent bactericidal agent. However, while it can react with most of the pathogens, it can also react with some of the dissolved substances that are present in the water, particularly the organic matter. This can lead to the formation of many other substances which are not normally present in food or water.

Analysis of chlorinated water has demonstrated a diversity of chlorination byproducts. Hundreds have been reported, the amounts and chemical structures varying according to the disinfection agent used, and the source of water [56]. The best known chlorination products are what are termed "halomethanes" (substances containing either chlorine, bromine or fluorine, and a carbon atom), in particular chloroform (one carbon and three chlorine atoms) and a related substance, bromoform, which is similar in structure but contains another halogen, bromine, instead. In addition, chlorination can also generate haloacetic acids, for example chloroacetic, dichloroacetic, and trichloroacetic acids ie acidic organic substances with two carbon atoms and one, two or three chlorine atoms. These different substances are all formed from the organic matter present in treated water. The combined total amount of these halomethanes and haloacetic acids is rarely in excess of 100 micrograms per litre ie 100 millionths of a gram per litre. However, the interaction of chlorine with what is mostly a mixture of organic substances that may be present in the water is highly complex as is clear from the number and diversity of chlorination products detected.

Many of the substances produced by chlorination probably rarely occur naturally in the environment. However, the addition of a chlorine atom or atoms to a chemical often appears to increase its environmental stability which really means that the normal degradative processes are less efficient. These processes may include degradation by microorganisms ie bacteria and fungi, or natural processes that include the effects of heat, light, water, atmospheric gases such as oxygen and carbon dioxide, and natural soil and rock constituents. Chemical stability may be an important quality if a chemical is to be used in certain situations, for example if there is a requirement for a chemical to be able to withstand excessive heat, acid, oxygen, or microbial degradation. However, this stability is a two edged sword because if chemically stable substances are released into the environment, their breakdown may be slower and this may result in their accumulation in the soil, water, and in animals, plants, and even humans. Examples of halogenated substances which show environmental stability include the organochlorine pesticides such as DDT, the artificial sweetener sucralose, and industrial chemicals such as polychlorinated biphenyls (PCBs) and polybrominated biphenyls (PBDEs). While the major water chlorination products, such as the halomethanes are degraded by the liver and excreted, little is really known about how well the great number of other disinfection byproducts, ie chlorinated, brominated, as well as non chlorinated or non brominated derivatives, are handled by our bodies.

Other Disinfection Processes

Because of the number and variety of chlorination products formed, and their potential toxicity, other procedures, termed advanced oxidation processes (AOPs), have been developed. These do not involve the use of chlorine and hence do not result in products containing chlorine atoms. Treatment with ozone, a gas whose molecular structure consists of three oxygen atoms (oxygen gas has two oxygen atoms), or hydrogen peroxide, and even ultraviolet light have been used [57]. These processes are being considered as alternatives to chlorination and the resulting generation of chlorinated products. However the use of these methods can also generate a variety of non chlorinated byproducts which differ in chemical structure from those formed from chlorination [30]. Like the chlorination products, these also are either not normally present in our food and water, or present in much smaller amounts. There is an underlying assumption that we can cope with whatever is formed.

Transport and Distribution of Drinking Water

After disinfection water has to then be transported to consumers via an elaborate system of holding reservoirs and underground pipes or water mains [58]. Once the water reaches a particular home, office or factory, it has then to be sent via pipes to different sites within each building. Mains and household transport involve the passage through pipes made from a variety of materials including reinforced concrete, plastics, metals such as steel, iron, copper or even lead, and asbestos cement. In addition, to reduce corrosion of some surfaces eg iron, coatings, such as bitumen or coal tar may used. Depending on the chemical structure of the piping, coatings used, its age, and some of the chemical characteristics of the water passing through the pipes eg whether the water is acidic, there may be release of components from the piping into the water [59–63]. For example, significant levels of lead have been reported to be released from lead piping into tap water, polycyclic hydrocarbons from pipes coated with coal tar, and even asbestos, released from asbestos cement. In addition to the release of chemical contaminants, faulty plumbing systems may also contain microorganisms which can contribute to gastrointestinal disease [63]. It is clear that, in addition to the disinfection products and other substances present after treatment by the water authority, traces of other contaminants may be introduced into water during its transport to the consumers.

Government Regulation of Drinking Water Quality

Water is such an important resource it is not surprising that there are guidelines for ensuring that drinking water is safe to drink and free of microorganisms and chemicals which could adversely affect the health of the community. In Australia these guidelines were developed by the National Health and Medical Research Council, a body established to develop and maintain health standards [52]. The NH and MRC is also the peak funding body for medical research in Australia. Not surprisingly given the importance of water quality, guidelines have also been developed by the World Health Organization, European Union, the USA EPA and other countries. While the Australian guidelines stress the critical importance of careful monitoring for the presence of microorganisms such as bacteria, protozoa, and viruses, ensuring that drinking water is largely free of chemical contaminants such as pesticides, industrial chemicals, and radioactive chemicals is also considered important. The guidelines provide a fairly comprehensive list of potential chemical contaminants as well as the maximum permitted levels of individual chemicals. Two hundred or more of possible contaminants in drinking water are listed in the NH and MRC water quality guidelines and include industrial chemicals and pollutants, insecticides (organochlorine, organophosphates as well as others), herbicides, metals such as arsenic, lead, copper and chromium, water chlorination derivatives, and even asbestos. In addition to the various chemical contaminants, the guidelines also include radioactive chemicals, both naturally occurring, for example uranium and thorium, and manufactured radioactive chemicals used in medicine and industry. Of course it is worth mentioning that the number and variety of chemicals that are used in agriculture and industry, some of which can find their way via their release into the air, waterways, and landfills, is immense and it is clearly exceedingly difficult to detect and measure everything. In addition to the NH and MRC guidelines there are a number of other Acts and regulations, both Federal and State, which relate to water quality and use. Some of these include the Safe Drinking Act (2003), the Health Fluoridation Act (1973), the Water Efficiency Labelling and Standards Act (2006), Safe Drinking Water Regulations (2005), and Environmental Protection Act.

Monitoring of Water Quality

It is fairly obvious that the composition and the levels of the various chemical contaminants in water can vary greatly according to many factors. The guidelines developed by government organizations such as the National Health and Medical Research Council provide for limits in the amounts of the various contaminants which means, in theory at least, that testing for contaminants must take place on a regular basis. Many of these contaminants have been mentioned earlier. However, the contaminants mostly analysed include metals such as copper, zinc, aluminium, arsenic, lead, chlorine, major disinfection products such as monochloroacetic, dichloroacetic, trichloroacetic acid, halomethanes, and inorganic components, for example nitrite and nitrate. Depending on the contaminant, analysis takes place mostly monthly or yearly. Analysis for many other numerous potential contaminants such as, for example, pesticides and industrial chemicals is rarely carried out. Despite this, numerous contaminants such as pesticides have been found in water at amounts greater than the NH and MRC limits. The Friends of the Earth have created an Australian Pesticide Map showing the places around Australia where the levels of pesticides detected in the water supply are above the limits imposed by the guidelines [64, 65]. It is important to note that there are a great number of insecticide and fungicide preparations used in agriculture. For example a search of the Australian Pesticide and Veterinary Medicine site shows that more than 1000 insecticide and fungicide preparations are used in Australian agriculture. Most of these preparations are formulations and they often contain other chemicals such as synergists, adjuvants, spreaders, and stabilisers. It is likely that at least some of these formulation chemicals may also be present in water although they, nor any disinfection products formed from them, are rarely measured.

Drinking Water and Disease

While water is essential to all life on the planet, including humankind, its ability to dissolve potentially toxic substances, and to contain a variety of microbes means that it is also a potential source of disease. There are numerous reports of disease outbreaks caused by bacterial, fungal, and viral contamination of water. Disease outbreaks caused by a variety of different bacteria including Cryptospyridium, Campylobacter, Shigella, and Giardia have been reported. Similarly, outbreaks of disease due to norovirus and rotavirus, are also well known [66]. In addition to microbial causes of disease, a variety of naturally occurring chemical contaminants have been shown to cause disease and there are indications that some of the chemicals produced by disinfection of water may also increase the risk of disease.

Naturally Occurring Chemical Contaminants and Disease

In many parts of the world, in particular in less developed countries, contaminants in groundwater have been clearly shown to cause disease even though the amounts consumed on a daily basis are much lower than the toxic level of the contaminant. Perhaps the best example is arsenic which is found in groundwater in parts of India and Bangladesh. Studies have shown that while much of the arsenic consumed in these areas is in drinking water, arsenic present in water used for irrigation can be taken up by crops, in particular rice [67]. While the tolerable intake is thought to be less than 2 microgram per kilogram body weight per day or around 120 micrograms per 60 kg body weight, in parts of West Bengal the amount taken up in food and water is almost three times this figure (329 microgram). At these levels, skin abnormalities including Bowen's disease (a type of skin cancer), foot swelling, enlargement of the liver and spleen, and even gangrene have been described. While the toxic dose, ie the dose which causes death is around 600 micrograms per kg body weight (ie 36 mg per 60 kg body weight), long term exposure to 1/100th of this amount taken per day over periods of time can cause disease.

Cadmium is another metal which can find its way into water supplies. It is relatively toxic and 20–30 grams can cause death. However long term exposure of much smaller amounts in water and food, such as that which has been shown to occur in rice grown using cadmium-polluted water for irrigation, can cause kidney disease [68]. A good illustration of how little is required to increase the risk of disease is the recommendation of the Japanese government that rice grown in an area where levels of cadmium of greater than one part per million ie one microgram per gram of rice should be considered inedible because of the risk of kidney disease. The lifetime dose of cadmium required to cause disease is believed to be less than 4 grams.

Lead is another metal which, even is small amounts, and taken over long periods of time, may have affects on health, and particularly brain function. Children are especially vulnerable. The amounts present in drinking water are normally below those thought to be harmful. However, additional exposure from other sources, such as smelting, mining, soil, lead paints and certain alternative medicines, may lead to elevated blood levels and an increased risk of disease [69].

Aluminium is the most common metallic element in the Earth's crust and occurs naturally combined with other elements such as oxygen, silicon, fluorine, and sodium. It is found in significant amounts in water from reservoirs and drinking water. Some of the aluminium in drinking water is derived from the coagulants used in the treatment of raw water. There is some evidence that aluminium in drinking water, particularly in combination with fluorine which increases absorption of the metal, may increase the risk of Alzheimer's disease [70].

Industrial, Agricultural, and Other Chemicals in Water and Disease

The number, and variety of chemicals produced by humans, is truly staggering. For example, the Australian Inventory of Industrial Chemicals lists more than 40, 000 chemicals available for industrial use in Australia while USA EPA chemicals and products database (CPDat) includes 49,000. With so many chemicals used it would be indeed surprising if at least some did not find their way into drinking water. A recent analysis of water from Lake Malaren, a Swedish lake providing water for up to 20% of the Swedish population, has shown the presence of at least 46 different chemical contaminants including pharmaceuticals and industrial chemicals [71]. In another report, 15 different pesticides were detected in Dutch untreated water [72]. Even assuming the treatment of water removes some of the contaminants, because of the variety of chemical structures it is probable that at least some chemical residues would remain, albeit in very small amounts. This has been confirmed in a Chinese study [73]. While it is well known that some of these chemicals, for example pesticides, do cause disease in occupational and other settings, there really is little information on any possible links between exposure to the small amounts that may be present in drinking water and any particular disease. However, there is some evidence of a possible role for fluorocarbons such as PFOS, which have many uses including firefighting foams and fabric treatment, in drinking water and thyroid abnormalities [74–77]. Another group of pollutants, nitrates, present naturally in water but often increased due to application of fertilisers and animal manures in agriculture, are of concern because of their potential conversion to nitroso compounds, known carcinogens, which in turn may increase the risk of intestinal cancers and thyroid disease [78, 79].

Radioactivity in Water and Disease

Radioactive elements, such as thorium and uranium exist naturally in the environment in rocks and soil, while some are released by human activities such as mining, and agriculture. In addition, radioactive substances are manufactured for medical and industrial purposes [12]. Whatever their source, radioactive elements, including artificial radionuclides made by humans released into the environment can, and do, find their way into lakes, rivers, seas as well as animals living in these environments [80, 81]. According to the NH and MRC Water Quality Guidelines the global average dose of natural radiation exposure per person is 2.4 mSv per year and less than 10% comes food and water combined. The toxic dose, which varies according to the type of radioactivity but is around 8 Sv, more than 30, 000 times the normal total yearly exposure of most people to the radiation in water and food combined [12]. However, as the normal daily exposure of most of the population to radiation occurs over many years, not a single year, and as there are different types of radiation, it is exceedingly difficult to determine whether long term exposure does cause disease. Reports on the effects of radioactivity in drinking water are limited. However, in studies carried out in Germany, there were suggestions of a possible link between drinking water uranium levels and different forms of cancer [82, 83]. Whether this effect is entirely due to the possible toxic effects of the metal itself or the radioactivity associated with the different isotopic forms (U-234, U-235, and U-238) is unclear. Similarly, there are reports that uranium in drinking water and food may contribute to kidney and bone diseases [84]. Exposure to groundwater radon, a radioactive gas present in indoor air and groundwater has been reported to increase the risk of lung and stomach cancers [85].

Plastics in Water and Disease

While most of the larger plastic items are easily removed by filtration and screening, what are much more difficult to remove are the tiny particles of plastic, microplastics, and even smaller particles termed "nanoplastics". Some microplastics are generated for use in different products such as cosmetics, hand and facial cleansers, and pharmaceuticals, while many are derived by the natural breakdown of larger plastic objects on land and at sea [86]. Some of them find their way into rivers and lakes, and then into reservoirs. The amounts, shape, size, and composition of the particles present in both treated and untreated waters varies considerably. Water treatment removes much of the plastic although from up to 100–600 particles per litre may remain in tap water [87, 88]. At least 12 different plastics have been identified with the most common including polyethylene terephthalate (PET), polypropylene (PP) and polyethylene (PE). Some of these particles may contain remnants of chemicals added during the manufacture of plastics. In addition, there is increasing evidence that microplastics are able to adsorb certain pollutants such as metals (eg chromium, copper, lead) and even non metals [89, 90]. Their presence in water, food, and even the atmosphere is of concern because of the possible health implications [91, 92]. There is, at present, no clear indication of an effect on human health. Experiments with fish show some effects on the gut, liver, and brain, while studies with mice have also shown effects on the liver and the gut [90, 92]. However, these studies mostly involve the use of microplastics derived from different plastics eg polystyrene, polyethylene, and polyvinylchloride, free of the pollutants that are often bound to the particles. At present it is not known whether the presence of absorbed pollutants on microplastics can produce other unexpected effects in humans, particularly if the exposure occurs over long periods of time.

Plumbing Contaminants and Disease

Contaminants in drinking water can arise through their release from household plumbing carrying water. One particular contaminant known to be released from plumbing is lead. The amounts of lead released appear to be related to the age and composition of the plumbing system, and the chemical characteristics of the water [93]. The Flint water crisis showed clearly that the chemical properties of tapwater from different sources may not be the same because in this particular case a change in the source of the water supply (from Detroit to the Flint river) resulted in mains water that was more corrosive thereby increasing its interaction with household plumbing. This in turn led to the release of lead from the piping and a corresponding increase in blood levels [94, 95]. Although the actual toxic dose for lead and its compounds is in the hundreds of milligrams per kg body weight (toxic dose of lead acetate is more than 700 mg per kg body weight) [96], it is believed that levels as low as 10–20 micrograms per litre of drinking water taken over periods of time may affect brain development and the function of the heart and kidneys, and is also thought to increase the risk of spontaneous abortions in pregnancy, low birth weight and preterm births [94].

Disinfection Byproducts and Disease

Normal tap water differs from the water which occurs naturally in rivers and lakes in that it has been chemically treated to remove the risk of microbial infections. The treatment process mostly involves chlorination which in turn leads to the production of a great number of disinfection byproducts which are not normally present in untreated or "raw" water. It has generally been assumed that these byproducts are not harmful, at least in the amounts normally present. It is worthwhile looking at the reasons for this belief in greater detail.

How Do We Determine Toxicity of Disinfection Byproducts?

The determination of the potential of disinfection byproducts which are present in treated water to cause harm can be carried out in a number of ways. Perhaps the simplest and most basic way is to study the effect of a particular byproduct on cells derived from animals or humans. This is a simple procedure and involves taking tissue from either humans or animals and culturing ie growing the cells which make up the tissue in a dish. The chemical is then added and the effect on the cells provides an indication of its likely toxicity. One property that is of considerable interest to scientists is the ability of certain chemicals to change the structure of the chemical components of our genes ie DNA. This property, referred to as "genotoxicity" has implications for many diseases, and in particular cancer. Research has shown that some of the disinfection byproducts are able to interact with DNA [97, 98]. While this method is useful, and is relatively harmless there are factors in living animals or humans which differ greatly from those that take place in cells grown in culture eg the uptake of a chemical in the body, travel to different organs via the blood, detoxification in the liver, and elimination via the kidneys.

An alternative approach is to use living animals such as mice, rats, dogs or even monkeys. This provides a better way of assessing the likely toxicity in humans but there are often large differences in how individual animal species handle chemicals. Also, while it is relatively easy to determine a toxic dose – one measurement is the LD50 test which is the dose at which 50% of a group of animals is killed – most of the chemicals in water are present in non toxic amounts but exposure may occur over extended periods of time, in some cases even years. Long term exposure can be assessed in animals but again, animals are not humans whose diets, genetics, life spans, as well as other factors, are quite different. There is a further factor which adds another layer of complexity to any attempt to determine likely toxicity and that is synergism. This refers to the ability of some chemicals to interact with others to produce unexpected effects. This concept has been discussed in another chapter in this book. What this means is that while individual chemicals in water have been shown to have no apparent effects at the concentrations present in drinking water, at least when tested in animals, what is not known is whether the mixture of chemicals present in disinfected water taken daily over long periods of time is also non toxic or whether there are some toxic effects caused by synergism between some of the components.

There is no doubt that some of the chlorination byproducts, such as the halomethanes are toxic if the dose is high enough. For example, toxicity studies in rats, one of the most commonly used animal species in toxicology studies, showed an LD50 value of 3–5 grams per kilogram body weight) ie the amount required to kill more than 50% of all animals tested. However, as the dose of these chemicals in water is so low, it is unlikely that the amounts of any of the contaminants even in a 24 h period is anywhere near the LD50 values. From studies with animals, chronic exposure ie exposure of a chemical over long periods of time may produce toxic effects even though the dose is much lower than the LD50 value. Good examples of this in humans include the exposure to the metals arsenic (mentioned earlier), cadmium, and lead. As we are not exposed to large amounts ie in the grams per kilogram range, a more relevant test which is carried out on animals is the NOAEL ie the no observed adverse effect level or the level at which no discernible effect is observed.

To illustrate the complexities of the determination of likely toxicity, it is worth looking at one of the contaminants in drinking water, trichloroacetic, formed by the chlorination process. LD50 values for rats and mice have been reported to be 3.3 and 4.9 grams per kg body weight respectively. In long term studies, trichloroacetic acid was dissolved in water and given to mice daily for periods of up to 82 weeks in much smaller amounts daily than the toxic dose [99]. Under these conditions the NOAEL (no observed adverse effect level) observed in mice was 78 mg/kg body weight, less than 1/40th of the toxic dose. For rats, the NOAEL was found to be 32 mg/kg body weight. At concentrations above the NOAEL but less than the LD50 the main effects noted were liver abnormalities. This demonstrates clearly that abnormalities are produced at levels much lower than a toxic dose provided the administration is long term. It also shows that there are differences in amounts required to produce effects in individual animal species.

There is yet another factor, mentioned earlier in this book, and relates to possible synergistic effects ie where a mixture of chemicals produces an effect greater than that which would be expected from the sum of the individual chemicals. It is worth emphasizing that there are dozens of disinfection byproducts and other chemicals, natural, agricultural, and industrial, in drinking water, and little is known about the effects of these complex mixtures taken over long periods of time. All we really know is that the toxic doses for most of these individual chemicals are much greater than the amounts present in water over a 24 h period or even longer.

How Are We Exposed to the Disinfection Byproducts and Other Chemicals in Water?

It is fairly obvious that the main route of exposure is through the water we drink. According to the US National Academy of Sciences, Engineering and Medicine, the intake of water in its different forms should be around 3.7 litres for men and 2.7 litres for women although there are many factors that can affect the amount of water consumed [100]. A significant fraction is in food. Most foods, vegetarian and non vegetarian, contain substantial amounts of water, mostly in the 60–90% range. Perhaps 2 litres or more, is tap or bottled water and drinks such as coffee and tea. Assuming a consumption of 2 litres per day of drinking water, and based on figures taken from one of the water supply sites (Melbourne Water) the amounts of trichloroacetic acid, just one of the chlorination byproducts, consumed per day, may vary from 6–96 microgams, while the amounts of another disinfection product, dichloroacetic acid varies from less than 1 to 42 micrograms per day.

Disinfection byproducts, such as the trihalomethanes, are also present in a variety of foods, including meat, dairy, vegetables and other foods, as well as beverages, and it is believed that they are absorbed by the food through washing and cooking [101, 102]. The amounts are not insignificant and depend on the type of food and temperature of the water [101].

Of course the water we drink or is in food is not the only source of these contaminants because we are exposed through washing of food, cooking, hand washing, watering of crops, bathing and in leisure activities such as swimming. It is also important to remember that exposure to the disinfection products is not just through ingestion but also occurs via the skin and lungs [103, 104]. There is also another entry point rarely considered and that is via the olfactory nerve in the nasal passages and from there directly into the brain, bypassing the blood-brain barrier which normally controls uptake into the organ. At present there is limited information on the possible significance of this pathway in humans for disinfection byproducts such as halomethanes although there is considerable research into the development of pharmaceuticals which enter the brain via this route [105]. Finally, there is yet another variable and that is the effect of heat, during food and drink preparation, and on bathing and showering, on the chemical structures of the disinfection byproducts. Heat is well known to induce changes in many different chemicals and it would be surprising if the various disinfection byproducts were completely stable to heat [106].

Are Disinfection Byproducts and Other Contaminants Taken up into Our Bodies?

In order for water contaminants to be harmful they have to first be taken up into our bodies through water or food, or via our lungs, skin or even the nasal passages directly into the brain. As there are literally hundreds of potential contaminants in drinking water it is easier to focus on those contaminants, for example, the trihalomethanes that are known to be present in significant amounts because these are the ones that are most studied. If these are detected it could be expected that at least some of the potentially hundreds of other contaminants found in water, often in much smaller amounts, would also be absorbed. As whatever is taken up eventually finds its way into the blood and from there into urine, it is probably best to focus on whether trihalomethanes are found in blood and urine. In fact, there are numerous reports of the presence of trihalomethanes in blood and urine and indications are that drinking, showering, and bathing are the main sources of these chemicals [103]. Trihalomethanes are not the only major disinfection byproduct found in blood and urine in significant amounts because another group of disinfection byproducts, haloacetic acids, have also been detected [104]. It would be surprising if at least some of the great number of other disinfection byproducts are not taken up. Therefore there is no doubt that some of the contaminants in disinfected water can, and do, find their way into our bodies.

Is There any Evidence That Chlorination Byproducts Cause Disease in Humans?

Any association between the chlorination byproducts in water and disease is difficult to establish because of the reliance on epidemiological studies, mentioned elsewhere in this book. There are a great number of variables which have to be taken into account in these types of studies and these include factors such as age, sex, diet, duration of exposure, amounts and types of chlorination products consumed, the period of exposure and how the exposure occurred. As exposure to these chemicals can occur through drinking water and other drinks, food, washing, bathing, and even swimming, these too must be considered. And, of course, the disinfection products always occur as mixtures with other chemicals, such as metals, salts, and organic matter and little is known of possible interactions between the disinfection byproducts and other chemicals in water. As there are so many variables in these types of studies, and as there are many different disinfection byproducts, much of the research has focused on trying to establish a link between one or more of the major disinfection byproducts, for example the trihalomethanes, and the risk of disease. It is even more complicated than this because the trihalomethanes themselves are not single chemical substances but a mixture and the actual composition almost certainly varies with the water supply. With this caveat, it is worth looking at the evidence that what is in our drinking water may affect our health.

Occupational Exposure to Disinfection Byproducts and Disease

It is not possible to carry out experiments on humans to assess the likely impact of long term exposure to non toxic amounts of a particular chemical. However, there are real life situations where exposure to small amounts of a chemical occurs over an extended period of time. The best example is exposure in an occupational setting. Many different diseases have been associated with occupational exposures to chemicals including respiratory conditions such as asthma and chronic obstructive lung disease, skin conditions, and cancer. The development of cancers in different parts of the body is a particularly good example of how small amounts of chemicals can induce cancers in a variety of different organs providing the exposure is over long periods of time such as that which occurs at work [107]. One of the working environments in which there is exposure to disinfection byproducts is swimming pools. Because of the increased risk of disease transmitted from infected individuals the water in swimming pools is often disinfected with chlorine. It could be expected that this treatment would result in the formation of disinfection products, including trihalomethanes, although the actual composition of the disinfection byproducts probably differs considerably from that found in drinking water due to the possible release of cosmetics, sunscreens, and whatever else is applied to the skin by swimmers, as well as sweat, hair, and even urine, in larger amounts than those found in drinking water. Also, exposure to contaminants in pools occurs mainly through the skin and lungs rather than by ingestion. There is evidence of an increased risk of eye and respiratory effects in swimming pool personnel which means that skin and lung exposure to chemicals in swimming pool water over periods of time may be harmful, at least for some people [108–110]. There are also indications that, even in non-occupational settings, and in particular for very young children who may have a greater vulnerability to chemical exposure, there may be an increased risk of lung conditions [111].

Disinfection Byproducts and Disease

As some of the disinfection products have been found to be mutagenic, and as mutagenic chemicals have been found to cause changes in DNA which in turn can lead to cancer, it is not surprising that much of the research on the impact of water contaminants on health has focused on their potential to cause cancer [97, 98]. In a study carried out in the USA and published in 2018 it was concluded that as many as 100, 000 lifetime cancer cases in the USA were due to potentially carcinogenic tap water contaminants, specifically arsenic, disinfection byproducts, and radioactive chemicals [112]. For individual cancers, there is some evidence of a link between bladder cancer and disinfection byproducts in water [113]. Analysis of data from 28 different EU countries showed an apparent relationship between the levels of one of the major classes of disinfection byproducts, trihalomethanes, and bladder cancer [114]. There are also suggestions that trihalomethane levels in drinking water are associated with an increased risk of rectal and colon cancers [115]. It is important to note, however, that trihalomethanes are mixtures of different substances and their total amounts present are greater than many of the other disinfection byproducts which are more difficult to measure. In effect they act as chemical surrogates for the great number of disinfection byproducts that are formed by chlorination. It is therefore possible that some of these other products contribute to the increased risk of bladder cancer.

An increased risk of cancer may not be the only effect of disinfectant byproduct exposure because there is now some evidence that the unborn child may be vulnerable. For example, there are suggestions that disinfection byproducts may increase the risk of stillbirth as well as craniofacial, heart, and musculoskeletal defects in babies at birth [116–118]. There are indications as well that birth weight is reduced with exposure to disinfection byproducts [119].

What Does This Mean?

There is no doubt that water is essential for all life on our planet, the myriad chemical reactions occurring in all life forms, including humans, taking place in an aqueous environment. However, because of its unique properties, the water in rivers, lakes, groundwater, the ocean, and even rain contain chemical pollutants originating from natural processes as well as the many, and diverse human activities. Because of this, drinking water derived from any source is rarely free of chemical, and in many cases, microbial, and even microplastic, pollutants. Water treatment processes remove some of the pollutants but significant amounts remain and disinfection of water kills most of the microorganisms. However, procedures used, in particular the most common, chlorination, result in the formation of a large number of disinfection byproducts, the amounts and chemical structures varying according to water source, time of year, weather, and water treatment and disinfection processes employed. Small amounts of a multitude of disinfection byproducts remain in the water. It has generally been assumed by governments around the world that the small amounts present are unlikely to cause disease even though, in larger amounts, there is clear evidence of toxicity in laboratory animals and in human and animal cells in culture. However this belief may have to be re-evaluated because there is clear evidence that exposure to non toxic amounts of certain chemicals in groundwater, for example the metals arsenic and cadmium, has been shown to cause disease provided the exposure occurs over many years. There is a further factor rarely considered by governments and that is possible synergistic effects resulting from the interaction of some of the chemical pollutants present in treated water to produce toxic effects greater than those that could be expected from the small, and apparently non toxic, amounts of individual pollutants. There is some evidence for this from reports that exposure to chlorinated water in an occupational setting in swimming pools may increase the risk of eye and respiratory effects. Also, even in non occupational settings there have been suggestions over many years that there may be a relationship between exposure to some of the disinfection byproducts in water and an increased risk of bladder and rectal cancers in adults, and developmental abnormalities in very young children. Because of possible synergistic effects newer methods to assess the toxicity of disinfection products and other components in drinking water need to be developed. In addition, alternative methods for disinfection which do not involve chlorination processes need to be considered.

References

1. Garrow JS, James WPT, Ralph A (2000) Human nutrition and dietetics, 10th ed. P14
2. Australian NH and MRC. Nutrient reference values (2006) Water
3. WHO (2019) Microplastics in Drinking Water
4. Bhowmick S et al (2018) Arsenic in groundwater of West Bengal, India. A review of human health risks and assessment of possible intervention options. Sci Total Environ 612:148–169

5. Uzun H et al (2020) Two years of post-wildfire impacts on dissolved organic matter, nitrogen, and precursors of disinfection bv-products in California stream waters. Water Res 181:115891

6. Battaglin WA et al (2018) Pharmaceuticals, hormones, pesticides, and other bioactive contaminants in water, sediment, and tissue from Rocky Mountain National Park, 2012–2013. Sci Total Environ 643:651–673. https://doi.org/10.1016/j.scitotenv.2018.06.150. Epub 2018 Jun 26

7. Pompei CME et al (2019) Occurrenece of PPCPs in a Brazilian water reservoir and their removal efficiency by ecological filtration. Chemosphere 226:210–219

8. Kabore HA et al (2018) Worldwide drinking water occurrence and levels of newly-identified perfluoroalkyl and polyfluoroalkyl substances. Sci Total Environ 616–617:1089–1100

9. Mawhinney DB et al (2011) Artificial sweetener sucralose in in US drinking water systems. Environ Sci Technol 45:8716–8722

10. Lopez-Pacheco IY et al (2019) Anthropogenic contaminants of high concern: existence in water resources and their adverse effects. Sci Total Environ 690:1068–1088

11. Bexfield LM et al (2019) Hormones and pharmaceuticals in groundwater used as a source of drinking water across the United States. Environ Sci Technol 53:2950–2960

12. NH and MRC Australian Drinking Water Guidelines (2011) Radiological quality of drinking water. Chapter 7

13. NH and MRC (2011) Administrative report. Review of lanthanum fact sheet. Australian Drinking Water Guidelines

14. Reynolds KA et al (2008) Risk of waterborne illness via drinking water in the United States. Rev Environ Contam Toxicol 192:117–158

15. Vaughese EA et al (2018) Estimating virus occurrence using Bayesian modeling in multiple drinking water systems of the United States. Sci Total Environ 619–620:1330–1339

16. La Rosa G et al (2020) Coronavirus in water environments: occurrence, persistence and concentration methods. A scoping review. Water Res 179:115899

17. Barros MUG et al (2019) Environmental factors associated with toxic cyanobacterial blooms across 20 drinking water reservoirs in a semi-arid region of Brazil. Harmful Algae 86:128–137

18. Masoner JR et al (2019) Urban stormwater: an overlooked pathway of extensive mixed contaminants to surface and groundwaters in the United States. Environ Sci Technol 53:10070–10081

19. Burant A et al (2018) Trace organic contaminants in urban runoff. Association with urban land-use. Environ Pollut 242:2068–2077

20. Sievers JC, Londong J (2018) Characterization of domestic graywater and graywater solids. Water Sci Technol 77:1196–1203

21. Oteng-Peprah M et al (2018) Greywater characteristics, treatment systems, reuse strategies and user perceptions- a review. Water Air Soil Pollut 229:255

22. Bibby K, Peccia J (2013) Identification of viral pathogen diversity in sewage sludge by metagenome analysis. Environ Sci Technol 47:1945–1951

23. Orive G et al (2020) Early SARS-CoV-2 outbreak detection by sewage-based epidemiology. Sci Total Environ 732:1392–1398

24. McEneff G, Schmidt W, Quinn B. US EPA. Research 142. Pharmaceuticals in the aquatic environment 2015. https://www.epa.ie/pubs/reports/research/health/Research%20142%20Report%20FINAL.pdf

25. Sobrino-Figueroa A (2018) Toxic effect of commercial detergents on organisms from different trophic levels. Environ Sci Pollut Res 25:13283–13291

26. Oberlin A et al (2020) Detergent formulations with high water content and anti-redoposition water polymers. US Patent Number 10597611

27. Chubaka CE et al (2018) A review of roof harvested rainwater in Australia. J Environ Public Health 6471324

28. US FDA. Bottled water everywhere. Keeping it safe https://www.fda.gov/consumers/consumer-updates/bottled-water-everywhere-keeping-it-safe

29. Aneck-Hahn N et al (2018) Estrogenic activity, selected plasticizers and potential health risks associated with bottled water in South Africa. J Water Health 16:253–262
30. Mason SA et al (2018) Synthetic polymer contamination in bottled water. Front Chem 6:407
31. Obmann B et al (2018) Small-sized microplastics and pigmented particles in bottled mineral water. Water Res 141:307–316
32. Akhbarizadeh R et al (2020) Worldwide bottled water occurrence of emerging contaminants: a review of recent scientific literature. J Hazard Mater 392:122271
33. Ngoc CN et al (2018) The erosive potential of additive artificial flavouring in bottled waer. Gen Dent 66:46–51
34. Polyakova OV et al (2018) Priority and emerging pollutants in the Moscow rain. Sci Total Environ 645:1126–1134
35. Enyoh CE et al (2019) Airborne microplastics: a review study on method for analysis, occurrence, movement and risks. Environ Monitor Assess 191:688
36. Nagaraj V et al (2018) Review- Bacteria and their extracellular polymeric substances causing biofouling on seawater desalination membranes. J Environ Manag 223:586–589
37. Coutinho de Paula E, Amaral MCS (2017) Extending the life-cycle of reverse osmosis membranes: a review. Rev Waste Manag Res 35:456–470
38. Kenigsberg C et al (2020) The effect of long-term brine discharge from desalination plants on benthic foraminifera. PLoS One 15:e:0227589
39. Koren G et al (2018) Seawater desalination and removal of iodine: effect on thyroid function. J Water Health 16:472–475
40. Koren G et al (2017) Seawater desalination and serum magnesium concentrations in Israel. J Water Health 15:296–299
41. Rowell C et al (2015) Potential health impacts of consuming desalinated water. J Water Health 13:437–435
42. Riva F et al (2018) Monitoring emerging contaminants in the drinking water of Milan and assessment of the human risk. Int J Hyg Environ Health 221:451–457
43. NH and MRC Australian Drinking Water Guidelines (2011) Chemical quality of drinking water
44. Australian Guidelines for Water Recycling. Managed Aquifer Recharge. July 2009. J Water Health
45. Soeder DJ (2018) Groundwater quality and hydraulic fracturing: current understanding and science needs. Ground Water 56:852–858
46. Melbourne Water. Water treatment. https://www.melbournewater.com.au/water/water-facts-and-history/why-melbournes-water-tastes-great-tap/water-treatment
47. Chahal CH et al (2016) Pathogen and particle association in wastewater. Significance and implications for treatment and disinfection processes. Adv Appl Microbiol 97:63–119
48. Sharma B et al (2017) Agricultural utilization of biosolids: a review of potential effects and soil and plant grown. Waste Manag 64:117–132
49. Clarke BO, Smith SR (2011) Review of 'emerging' organic contaminants in biosolids and assessment of international research priorities for the agricultural use of biosolids. Environ Int 37:226–247
50. Meng XZ et al (2016) Organic contaminants in Chinese sewage sludge: a meta-analysis of the literature of the past 30 years. Meta-Anal Environ Sci Technol 50:5465–5466
51. Mohajerani A, Karabatak B (2020) Microplastics and pollutants in biosolids have contaminated agricultural soils. Rev Waste MANAG 107:252–265
52. NH and MRC Australian Drinking Water Guidelines (2011) 94–99
53. Behets GJ et al (2020) Use of lanthanum for water treatment: a matter of concern. Chemosphere 239:124780
54. Melbourne Water. Testing water quality. https://www.melbournewater.com.au/water-data-and-education/water-facts-and-history/why-melbournes-water-tastes-great/testing-water
55. Yoon GS et al (2018) Selection criteria for oxidation method in total organic carbon measurement. Chemosphere 199:453–458

56. Kimura SY et al (2019) The DB exposome. Development of a new method to simultaneously quantify priority disinfection by-products and comprehensively identify unknowns. Water Res 148:324–333

57. Kuo J Disinfection processes. Water Environ Res 89:1206–1244

58. National Academies of Science, Engineering and Medicine (1982) Elements of water supply. Drinking Water Health 4

59. Blokker E et al (2013) Health implications of PAH release from coated cast iron during drinking water distribution systems in the Netherlands. Environ Health Perspect 121:600–606

60. Rajasarkka J et al (2016) Drinking water contaminants from epoxy resin-coated pipes. A field study. Water Res 103, 133–140.38. Jarvis P et al (2018) Intake of lead (Pb) from tap water of homes systematic review and meta-analysis. Environ Health Perspect 122, 651–660

61. Saitoh K et al (1992) Concentration and form of asbestos fiber in tap drinking water contaminated from a water supply pipe with asbestos-cement. Nihon Eisegaku Zasshi 47:85–860

62. Browne ML et al (2005) Cancer incidence and asbestos in drinking water, Town of Woodstock, New York, 1980–1988. Environ Res 98, 224–232.with leaded and low lead plumbing systems. Sci Total Envi ron 644:1346–1356

63. Ercumen A et al (2014) Water distribution system deficiencies and gastrointestinal illness: a systematic review and meta-analysis. Environ Health Perspect 122:651–660

64. Friends of the Earth. Australian Pesticides Map. https://pesticides.australianmap.net/chemicals/

65. Amis A (2017) Under the radar Pesticide detections Victorian water supply. Friends Earth Chain Reaction 131:6

66. Moreira NA, Bondelind M (2017) Safe drinking water and waterborne outbreaks. J Water Health 15:83–96

67. Bhowmick S et al (2018) Arsenic in groundwater of West Bengal, India: a review of human health risks and assessment of possible intervention options. Sci Total Environ 612:148–169

68. Kobayashi E et al (2009) Influence of drinking and/or cooking with Jinzu River water on the development of Itai-Itai disease. Biol Trace Elem Res 129:46–57

69. Brown MJ, Margolis S (2012) Lead in drinking water and human blood levels in the United States. MMWR Suppl 61:1–9

70. Russ TC et al (2020) Aluminium and fluoride in drinking water in relation to later dementia risk. Br J Psychiatry 216:29–34

71. Rehrl AL et al (2020) Spatial and seasonal trends of organic micropollutants in Sweden's most important water reservoir. Chemosphere 249:126168

72. Sierps RMA et al (2019) Occurrence of pesticides in Dutch drinking water sources. Chemosphere 235:510–518

73. Wan Y et al (2019) Neonicotinoids in raw, finished and tap water from Wuhan, Central China: assessment of human exposure potential. Sci Total Environ 675:513–519

74. Kabore HA et al (2018) Worldwide drinking water occurrence and levels of newly-identified perfluoroalkyl and polyfluoroalkyl substances. Sci Total Environ 616–617:1089–1100

75. Winquist A, Steenland K (2014) Perfluorooctanoic acid exposure and thyroid disease in community and worker cohorts. Epidemiology 25:255–264

76. Andersson EM et al (2019) High exposure to perfluorinated compounds in drinking water and thyroid disease. A cohort study from Ronneby, Sweden. Environ Res 176:108540

77. Knox SS et al (2011) Perfluorocarbon exposure, gender and thyroid function in the C8 Healh project. J Toxicol Sci 36:403–410

78. Loh YH et al (2011) N-Nitroso compounds and cancer incidence: the European prospective investigation into cancer and nutrition (EPIC)-Norfolk study. Am J Clin Nutr 93:1053–1061

79. Ward MH et al (2018) Drinking water nitrate and human health: an updated review. Int J Environ Res Public Health 15(7):1557

80. Martinez J et al (2018) Presence of artificial radionuclides in samples from potable water and wastewater treatment plants. J Environ Radioact 192:187–193

81. Carvalho FP (2018) Radionuclide concentration processes in marine organisms. A comprehensive review. Rev J Environ Radioact 186:124–130
82. Banning A, Benfer M (2017) Drinking water uranium and potential health effects in the German Federal State of Bavaria. Int J Environ Res Public Health 14(8):927
83. Radespiel-Troger M, Meyer M (2013) Association between drinking water uranium content and cancer risk in Bavaria, Germany. Int Arch Occup Environ Health 86:767–776
84. Shin W et al (2016) Distribution and potential health risk of groundwater uranium in Korea. Chemosphere 163:108–115
85. Kyle P et al (2018) Lung and stomach cancer associations with groundwater radon in North Carolina, USA. Int J Epidemiol 46:676–685
86. Karbalaei S et al (2018) Occurrence, sources, human health impacts and mitigation of microplastic pollution. Environ Sci Pollut Res Int 25:36046–36063
87. Koelmans AA et al (2019) Microplastics in freshwaters and drinking water. Critical review and assessment of data quality. Water Res 155:410–422
88. Pivokonsky M et al (2018) Occurrence of microplastics in raw and treated drinking water. Sci Total Environ 643:1644–1651
89. Godoy V et al (2019) The potential of microplastics as carriers of metals. Environ Pollut 255:113363
90. Alimba CG, Faggio C (2019) Microplastics in the marine environment: current trends in environmental pollution and mechanisms of toxicological profile. Environ Toxicol Pharmacol 68:61–74
91. Toussaint B et al (2019) Review of micro- and nanoplastic contamination of the food chain. Food Addit Contam Part A Chem Anal Control Expo Risk Assess 36:639–673
92. Yong C et al (2020) Toxicity of microplastics and nanoplastics in mammalian systems. Int J Environ Res Public Health 17(5):1509
93. Jarvis P et al (2018) Intake of lead (Pb) from tap water of homes with leaded and low leaded plumbing systems. Sci Total Environ 644:1346–1356
94. Levallois P et al (2018) Public health consequences of lead in drinking water. Curr Environ Health Rep 5:255–262
95. Ruckart PZ et al (2019) The Flint water crisis: a coordinated public health emergency response and recovery initiative. J Public Health Manag Prac 25(Suppl 1)
96. Centres for Disease Control and Prevention. NIOSH. Lead compounds as Pb. https://www.cdc.gov/niosh/idlh/7439921.html
97. Richardson SD et al (2007) Occurrence, genotoxicity, and carcinogenicity of regulated and emerging disinfection products in drinking water: a review and roadmap for research. Mut Res 636:178–242
98. Cortes C, Marcos R (2018) Genotoxicity of disinfection byproducts and disinfected waters. A review of recent literature. Mutat Res Genet Toxicol Environ Mutagen 831:1–12
99. WHO guidelines for drinking water quality (2004) Trichloroacetic acid in drinking water. https://www.who.int/water_sanitation_health/water-quality/guidelines/chemicals/trichloro aceticacid.pdf
100. The National Academies of Sciences, Engineering and Medicine. Dietary Reference Intakes for Water, Potassium, Sodium, Chloride, and Sulfate. 2004. February 11, 2004
101. Huang AT, Batterman S (2010) Sorption of trihalomethanes in food. Environ Int 36:754–762
102. Cardador M et al (2016) Detection of regulated disinfection by-products in cheeses. Food Chem 204:306–313
103. Chowdhury IR et al (2020) Human exposure and risk of trihalomethanes during continuous showering events. Sci Total Environ 701:134521
104. Zhang W et al (2009) Reliability of using urinary and blood trichloroacetic acid as a biomarker of exposure to chlorinated drinking water disinfection byproducts. Biomarkers 14:355–365
105. Crowe TP et al (2018) Mechanism of intranasal drug delivery directly into the brain. Life Sci 195:44–52

106. Liu B, Reckhow DA (2015) Disparity in disinfection byproducts concentration between hot and cold tap water. Water Res 70:196–204
107. Cancer Council of Western Australia. Monograph Series (2015) Occupational exposures to carcinogens in Australia
108. Bureau G et al (2017) Indoor swimming pool environments and self-reported irritative and respiratory symptoms among lifeguards. Int J Environ Health Res 27:306–322
109. Kanikowska A et al (2018) Influence of chlorinated water on the development of allergic diseases-an overview. Ann Agric Environ Med 25:651–655
110. Chiu SK et al (2017) Respiratory and ocular symptoms among employees of an indoor waterpark resort-Ohio, 2016. MMWR Morb Mortal Wkly Rep 66:9860989
111. Andersson M et al (2018) Early life swimming pool exposure and asthma onset in children- a case control study. Environ Health 17:34
112. Evans S et al (2019) Cumulative risk analysis of carcinogenic contaminants in United States drinking water. Heliyon 5, (9)e02314)
113. Diana M et al (2019) Disinfection byproducts potentially responsible for the association between chlorinated drinking water and bladder cancer: a review. Water Res 162:492–504
114. Evlampidou I et al (2020) Trihalomethanes in drinking water and bladder cancer burden in the European Union. Environ Health Perspect 128(1):17001
115. Jones RR et al (2019) Ingested nitrate, disinfection by-products, and risk of colon and rectal cancers in the Iowa Women's study cohort. Environ Int 126:247–251
116. Rivera-Nunez Z et al (2018) Exposure to disinfectant by-products and risk of stillbirth in Massachusetts. Occup Environ Med 75:742–751
117. Wright JM et al (2017) Disinfection by-product exposures and risk of specific cardiac birth defects. Environ Health Perspect 125:269–277
118. Kaufman JA et al (2020) Disinfection by-product exposures and risk of musculoskeletal birth defects. Environ Epidemiol 4(1)
119. Save-Soderbergh M et al (2020) Exposure to drinking water chlorination by-products and fetal growth and prematurity A nationwide register-based prospective study. Environ Health Perspect 128:57006

Chapter 8
Paper Manufacture and Use

The advent of computers was thought to increase our reliance on the electronic word and this in turn was believed to reduce our need for the printed word and for paper. In other words, what was envisaged was a paperless society. With the benefit of hindsight we can see that we were somewhat naïve because, to use the jargon, computer "soft copies" have not eliminated the need for "hard copies" but rather may have even increased the use of paper. And, of course, there are many other applications of paper – from food packaging, to transport of goods, newspapers, books, labeling, cards, stationery, tissues, and toilet paper to name just a few. The total amount of paper used in the world, in its many applications, is around 300 million tons. The figure cited for the USA, 90, 000 tons, indicates that each person in the USA uses about 700 pounds per year, a truly staggering figure. It is even more amazing considering that the population of this most technologically advanced country is only 300 million, or 5% of the world population. Thus, on a per capita basis, the USA uses six times the amount of paper used by the rest of the world clearly demonstrating that technological advance does not necessarily lead to so-called "paperless society".

Like everything else made by humans, raw materials are required, and some sort of industrial process is necessary to convert the raw material – which in this case is the wood from trees – to the finished product. And, as we have also seen before, because of its myriad uses, it would be surprising if there was not some sort of chemical treatment to generate a product with desirable attributes. Once the product is made and used it has to be disposed of and, in the case of paper, either by recycling, burning or as landfill. Chemical agents are used in the manufacture and in modification of paper, and they, or products derived from these agents, eventually find their way into the environment. And, when paper degrades, either by burning or within a landfill, there is a further release of chemicals into the environment. While the amount generated per kilo or pound of paper manufactured may not be large, because of the vast amount of paper produced and used it becomes significant. And, like the chemical by products generated by other industrial processes and

© The Author(s), under exclusive license to Springer Nature Switzerland AG 2021
A. Poulos, *The Secret Life of Chemicals*,
https://doi.org/10.1007/978-3-030-80338-4_8

degradation, it is obvious that there is a potential for at least some of these chemicals to find their way into our bodies, either through our food, water, or the air we breathe.

In the past, government regulation of this release of chemicals was perhaps more lax than it is today. Despite the improvements in government regulation today, there are enormous pressures on governments to approve pulp and paper manufacture projects, not least because of the benefits in employment. The argument used by the industry has always been that the types of chemical released, and their quantity, is insufficient to cause harm either to the environment or to humans. Whilst this may or may not be the case, there are plenty of examples where the use of certain chemicals, originally approved by the government, has been discontinued due to new evidence showing the possibility of harm. Indeed, this has already occurred in pulp manufacturing with the substitution for chlorine dioxide for chlorine in bleaching. In other words, governments are not infallible, particularly where there is political pressure. The fact is that a diverse array of chemicals is released into the environment during the manufacture of paper, and others are released during recycling and waste disposal. When the varieties of paper ie treated paper, and the amounts of paper used, were relatively limited, it was possible to argue that environmental damage was not very great or at least acceptable. However, with the exponential increase in the use of paper for purposes undreamed of a hundred years ago, it is more difficult to argue that there is no environmental impact. There are quite genuine concerns that some of the chemical by products of paper manufacture and use are potentially harmful. But are these concerns justified? What are the chemical by products, and is there any evidence that they may be harmful? In this chapter we will address some of these questions.

Chemical Byproducts of Paper Manufacture

Before we discuss the chemical byproducts we need to know a little of the process. There is some controversy as to when paper was first invented. However, there is little doubt that a type of paper, made from papyrus, was in use in ancient Egypt more than 4000 years ago. It is believed that true paper was invented in China, perhaps as early as the second century AD, and gradually spread around the world until the relatively recent development of mechanized production in the nineteenth century.

The paper that we use today is essentially composed of compressed cellulose fibres derived from wood. Cellulose is a complex sugar, ie it is a polymer of the sugar, glucose. Each cellulose molecule is made up of thousands of glucose molecules, joined together like a string of beads where each bead represents a glucose molecule. The cell wall of plants, a type of skeleton holding the cell together, is made up mainly of cellulose. We are all familiar with starch which, like cellulose, is also a polymer of glucose but differs in the way the individual glucose molecules are joined together. Whereas humans are able to degrade starch in the intestine, cellulose is much more resistant and not degraded. Higher grade paper is made by removing certain chemical substances that are linked to cellulose. The principal substance removed is lignin which is also a polymer but not of glucose but of three even more complicated substances with rather long winded names – paracoumaryl alcohol, coniferyl alcohol, and sinapyl alcohol [6]. Paper can be made by leaving much of the lignin but it tends to deteriorate with age due to the breakdown of lignin.

Paper manufacture generally involves the formation of wood pulp, a process is carried out in 'pulp mills'. If lignin is to be removed, the pulp is treated with acidic materials such as sulphite (formed by dissolving the gas sulphur dioxide in water) or via the Kraft process that utilizes non-acidic materials (sodium hydroxide and sodium sulphide in water). High temperatures over extended periods are used in both processes and generate cellulose largely devoid of lignins or any other substances present in wood. However, at the high temperatures, and in the presence of reactive substances such as sulphite and sodium hydroxide, it would be indeed surprising if all that happened was the release of lignin into the aqueous medium. The resulting mixture contains cellulose fibres and lignin together with a very complex mixture of chemical substances dissolved in water. These substances are either derived from lignin itself or represent chemically modified forms of lignin, but also contain a great number of other constituents of wood. It is important to note that many of the substances present are not naturally occurring because the conditions that are used to make paper are quite different from those occurring in plants. Once the cellulose is removed to make paper, most of what is left is a waste product of the process.

Bleaching of Pulp

Because of the requirement for white paper, the pulp, which is mostly a dark colour, is bleached. In the past, chlorine was used although, because of concerns about the toxicity of substances produced by this process, chlorine dioxide (which also contains chlorine) has been increasingly used. Attempts have been made to use alternatives free of chlorine completely, termed "totally chlorine free" or TCF, such as ozone or hydrogen peroxide, but these have not completely replaced the use of chlorine dioxide. Unfortunately, while chlorination is a highly effective bleaching method, it comes at a cost – a variety of chlorinated substances are formed because of the very high chemical reactivity of chlorine or chlorine dioxide [29]. We will discuss this later.

Non Chemical Processes

Because of concerns about the generation of highly coloured and potentially toxic substances during the production and bleaching of pulp, alternative methods reliant on biological systems have been developed. At present little is known about the environmental impacts of these processes [7, 26].

What Else Is Added to Paper?

The bleached pulp forms the basis for the production of different types of paper. There is an incredible variety of paper used, each with its own particular function. These include, for example, greaseproof and baking papers, newsprint, copy paper, cardboard, bacon paper, fluorescent paper, ammunition paper, granite paper, thermal paper, and glazed paper to name just a few. A variety of chemical substances may be added at some point during the manufacturing process to produce a paper with the desired qualities. For example, the production of carbonless copy paper involves the use of plastics, solvents, dyes, adhesives, defoaming agents, ultraviolet light absorbers etc. [15].

What Are the Potential Environmental Contaminants Released During the Manufacture and Disposal of Paper?

The paper industry uses large amounts of water during the manufacture of paper. It has been estimated that the production of one tonne of paper requires up to 450 cubic metres of water, and up to 300 cubic metres of waste water, or effluent containing a great variety of potentially toxic substances and intensely coloured effluents are produced [7]. Discharge of these effluents causes severe water pollution.

As mentioned above, chemicals are added at various stages of paper manufacture. During the formation of wood pulp, anthraquinone, defoamers, pigments, fillers, surfactants (ie soaps), and chelators (substances that bind metals) to remove metals etc. added to improve the yield of pulp and to assist in the removal of lignin [17]. Almost certainly, some of these find their way into the environment despite the increasingly stringent procedures to limit their release. The treatment of pulp also generates many other substances which are normal components of wood. Some of these materials are almost certainly changed chemically during the removal of lignin or bleaching. Indeed, because of the complexity of the chemical processes used, it is unlikely that everything chemically produced has been identified. We know, for example, that substances are produced from lignin, but wood, like all other plant-derived materials is a very rich source of chemicals other than lignin. In a recent study of kraft pulp mill waste, more than 30 different substances were positively identified but it is clear that this is the tip of the iceberg [1, 8].

While chemicals are produced at all stages in the paper manufacturing process, the area of greatest concern as regards the generation of environmental contaminants is the bleaching of pulp and in particular the use of chlorine as a bleaching agent because of its ability to form environmentally stable toxins such as dioxins [8, 18, 23]. In addition to the chlorinated and non-chlorinated substances formed, there is a great variety of others, including gases such as carbon monoxide, ammonia, and nitrogen oxide, as well as sulphur-containing gases such as hydrogen sulphide, methyl mercaptan, dimethyl sulphide, and sulphur dioxide. Most of these gases are toxic. Sulphur dioxide is a major pollutant released from pulp mills and, because of its ability to react with water to form an acid, sulphurous acid, contributes to acid rain.

In addition to pollutants generated from the manufacture of paper, there is further potential for pollution through the disposal of paper and associated products such as cardboard. The principal methods of waste disposal of paper include landfill, incineration and recycling.

What Impact Does Paper Manufacture and Disposal Have on Our Health and Wellbeing?

Undoubtedly the aspect of paper manufacture and use that has generated the most controversy in relation to possible effects on our health is pulp manufacture and processing. In particular, the use of chlorine as a bleaching agent for pulp has been widely criticized because of the generation of environmentally stable chlorinated substances and has led to the development of alternative bleaching procedures.

It might be useful to explain why chlorine is used for bleaching and why it has the potential to generate unwanted products. Chlorine is a very nasty greenish gas with a suffocating odour. As it makes up more than 50% of common salt, it is abundant and relatively cheaply produced from this source. When it combines chemically with water it produces something called "hypochlorous acid" which has particularly useful properties as an antiseptic. Because of its considerable chemical reactivity, microorganisms in water are killed by hypochlorous acid – hence its use to sterilize drinking water and swimming pools. It has a second property which is particularly useful in bleaching, which is essentially the chemical removal of pigmented material. Most pigments that occur in plants, and this includes the pigments in wood pulp, exist as carbon containing (ie organic) substances. These substances, as well as many other non pigmented organic substances, interact with chlorine (or strictly speaking hypochlorous acid) through a process termed "oxidation". The chemical structure of the pigment is thereby changed resulting in the loss of colour.

Unfortunately, as is the case in the sterilization of water, this comes at a price – the generation of a potentially great number of "chlorinated" substances. Essentially what may happen is that one or more chlorine atoms is added to some of the organic substances that are present in water. When water is sterilized with chlorine for drinking purposes, a number of chlorinated substances, including the trihalomethanes (THMs) are formed, some of which are considered potentially harmful. Of course, there is a world of difference between bleaching or sterilizing water for drinking and bleaching wood pulp. Generally speaking, in contrast to wood pulp, drinking water contains relatively small amounts of organic material so the corresponding amounts of chlorinated materials formed will be small. Moreover, the types of organic matter in drinking water and wood pulp are very different and hence any chlorinated products formed are also likely to be different.

In addition to the fact that the addition of a chlorine has the potential at least of generating something that is more harmful than the corresponding non-chlorinated substance, there is an added problem. And that is that the addition of a chlorine atom to a substance often increases the environmental stability. That is, the natural processes that degrade some environmental pollutants only work very slowly and so they can build up in the soil, rivers and the ocean. In fact that is exactly what has happened in the case of at least two classes of chlorinated substances released into the environment – the antimalarial DDT and certain chlorinated pesticides.

There have been numerous reports over the last 20 years or so on the effects of paper mill effluents on the reproduction of many different fish species. The most common effects include a reduction in the size of the sexual organs, a reduction in the blood levels of sex hormones, and a decrease in the numbers of eggs produced [9]. It is believed that these effects are produced through the ability of certain substances in the effluents to interfere with the effects of steroid hormones on normal developmental processes. The present view is that these substances, which have not been fully identified, but are collectively referred to as "endocrine disrupting substances" are able to interact with certain proteins, termed "receptors". The word endocrine refers to certain glands such as the adrenals, pituitary, thyroid and pancreas to list just a few which produce hormones – they may be proteins (eg insulin produced by the pancreas) or non proteins like steroids (made by sexual organs like the ovaries or testes) or adrenaline (made by the adrenals)- and then release them into the blood or lymph where they travel throughout the body and effect the function of a particular organ. The release of adrenaline from the adrenals that affects the function of the heart is a good illustration. All hormones have a receptor which is present in one or more different tissues in the body. Steroid hormones produce their effects through their interaction with its specific steroid hormone receptor so that anything that affects this will affect reproduction as well as growth and development.

While the substitution of other bleaching agents for chlorine reduced the effects, it has been concluded that the effluents from all pulp mills are potentially harmful to fish [3]. Of course skeptics would argue that just because a few fish have reproductive anomalies is not necessarily relevant to humans. After all, so the argument goes, the levels of chemical exposure of fish (particularly those that live near pulp mill effluents) are much higher than the exposure of humans. While it may be true, whatever is released into rivers can find its way into the sea and from there into marine animals and then into the food chain of humans. There are too many examples (mercury, PCBs, PBDE and pesticides are just a few) for there to be any doubt about this. Moreover, some products generated by kraft pulp mills have already been detected in fish tissues and it would be surprising if many more pulp mill chemicals are not taken up into fish [32]. Indeed, the effects of the pulp mill chemicals on the reproduction of fish would tend to suggest this. It is also worth emphasizing also that many of the processes that occur in fish, in particular the regulation of reproduction through the interaction of hormones with their respective receptors, also occur in humans. It is therefore true that whatever happens in animals is a wake up call for humans.

Incineration of Paper and Pollutants

Pulp mills are not the only source of pollutants that may impact on our health and wellbeing. The disposal of paper, particularly through incineration can generate other harmful substances [19, 30]. Small amounts of dioxins are produced by burning of paper, the amount depending on the type of paper burnt [14]. Treatment of paper, often used to make paper for specific applications, can greatly increase the yield of pollutants [22, 30]. Studies have demonstrated that exposure of rats to TCDD, perhaps the most studied of the number of dioxins that may be formed by natural (eg bushfires) and human activities, is very poisonous, the LD50 (a measure of the amount required to kill more than half of a group of animals) varying from as little as a half a millionth of a gram in guinea pigs to around 40 millionths of a gram in rats, and 70 millionths of a gram in rhesus monkeys. In addition to its acute effects, if administered in amounts smaller than that required to cause death TCDD over periods of time it has been shown to cause cancers of the liver, lung and mouth in rats [28]. There are indications as well that TCDD can also affect heart function [12].

As far as humans are concerned, we know the effects that can be produced from accidental release of dioxins into the environment. The best known example of this is the explosion that occurred in 1976 in Seveso, a populated area to the north of Milan in Italy. An explosion in a factory that manufactured a precursor of the herbicide 2, 4, 5 T resulted in the release of around 1 kg of TCDD (tetrachloro dibenzodioxin is the chemical name) into the environment. Numerous studies have been carried out on the population exposed to TCDD. A comprehensive study involving a 25 year follow up on the effects on 278,000 individuals exposed to the toxin was published recently [4]. The most documented effect is in the development of chloroacne, a severe form of acne which has also been observed on exposure to other related substances such as PCBs. Longer term effects included cancers of the lymph (fluid from the blood which travels through the small blood vessels to nourish tissues and then is transported back into the blood via lymph vessels that are analogous to blood vessels), heart disease, and diabetes. The suggested association between TCDD exposure and diabetes, heart disease, and cancer is not confined to the population living in or near Seveso but has also been reported in people around the world whose exposure has been much more limited presumably through their occupation, their proximity to waste incinerators, or in food [11, 24, 25]. The latter, particularly fatty fish and fatty meats are significant sources of dioxins but there are indications that the levels are dropping. It should be noted that dioxins are also formed during the bleaching process, particularly if chlorine is used. Effluents, both treated and untreated, contain dioxins although the amounts formed are reduced greatly if chlorine dioxide is substituted for chlorine [23].

Landills and Pollutants

In relation to landfill, unlike plastic, the breakdown of paper in landfill is relatively straight forward. Cellulose, the main component of paper, is degraded fairly rapidly by soil microorganisms, the ultimate product being the simple sugar glucose which is non-toxic. However, depending on its use, the other substances that have been added to the paper and mentioned above, such as inks, pigments, fillers, plastic, silicon linings, adhesives, preservatives etc., are released into the soil. There are even reports of the presence of dioxins in paper so presumably they too will be released as will any residual lignins and products derived from lignins that have been formed during bleaching [31]. If it is a carbon containing substance it may be acted upon by microorganisms and, either degraded completely into simple substances such as, for example carbon dioxide, methane, and water, or it can be modified chemically in some other way. The organic modified or non-modified forms, or any non-organic additives such as barium, calcium, or magnesium salts, may find their way into a river, perhaps into the water supply and possibly the ocean. With the vast amounts of paper sitting in landfills, even trace amounts of contaminants will add up and leave their chemical footprint on the environment. Unfortunately, little is really known of what is produced from the breakdown of paper in landfills.

Recycling and Pollutants

The recycling of paper generates its own chemicals including estrogenic substances like bisphenol A and other related substances [5, 21]. Bisphenol A is a member of the group of endocrine disrupting substances already referred to above. It is a component of certain papers, in particular thermal papers. There has been lots of research into the effects of bisphenol A. While it clear that, in a test tube, bisphenol A can bind to certain receptors, in particular certain hormone receptors, it is not entirely clear whether the small amounts of bisphenol A to which we are probably all exposed are harmful to our health. However, in a paper published in 2005, it was concluded that, based on a review of 115 reports of other researchers published over a number of years, there were ample evidence that the concentrations of bisphenol A detected in human blood and tissues, as well as the blood of fetuses, was sufficient to cause behavioral abnormalities in mice [16, 27]. This particular report was highly critical of another report, funded by American Plastics Council, which concluded that low doses produced little measurable effects. There is some evidence that exposure, as determined by the measurement of the levels of bisphenol A in the urine, is linked to degenerative diseases such as diabetes and cardiovascular diseases, as well as abnormalities in liver function and obesity and abnormalities of the ovaries in women [13, 20]. On the other hand, the US government agency, the FDA, believes that "the scientific evidence at this time does not suggest that the very low levels of human exposure to BPA through the diet are unsafe" [10]. The FDA report has in turn been criticized for its preference for industry-funded rather than government funded research, the argument being that industry-funded research may be biased [14].

In addition to bisphenol A a recent study demonstrated the presence of more than 250 substances in recycled paperboard [2]. The number and nature of these substances seemed to vary according to the source of the cardboard. Many of these were identified as arising from the printing inks, from the products used for making paper and paperboard, and chemicals added to paper. As one of the major uses of paperboard is in the manufacture of food packaging materials, there were concerns about possible migration of many of these substances into the packaged food.

What Does All This Mean?

There is no question that the volume of paper manufactured throughout the world has increased markedly over the last decade or so despite the earlier predictions of a paperless society. As in may other spheres of human industrial activity there is, predictably, a parallel increase in the use, and subsequent environmental release, of a great number of chemical substances, many of which are not normally present in the environment or, if they are, in much smaller amounts. And, again, it has been assumed that whatever is released is not harmful either to animals or humans. Now as far as the chemicals released as by products of paper manufacture are concerned we know that this is clearly not the case. We have known for a long time that some of the chemicals that are released into the waterways can affect such basic processes in fish as reproduction but despite this, it is really only relatively recently that one of the processes in paper manufacture that is believed to generate chemicals that may affect these processes ie chlorination of pulp has been gradually phased out around the world. Of course it is not just chlorination that has the capacity to produce harmful chemicals because even with the newer bleaching processes that have been developed it is really too early to be sure of the environmental and human impact. There is little doubt that non chlorine bleaching also generates an extraordinary variety of chemical pollutants. And pulp manufacture is not the sole source of pollutants because recycling and incineration also have been reported to generate harmful substances eg bisphenol A and dioxins., and we know very little about the long term effects of release of yet other material from landfills. Whatever the human activity, the pattern of behaviour is the same. A technology, such as chlorination, is embraced eagerly at first but then, over time, as reports of possible environmental harm begin to grow, there is initial dismissal of the relevance of the reports but, gradually, as the evidence grows stronger, there is a gradual acceptance that there may be something in it leading to a gradual phasing out of the technology. While this is going on, the many years may have elapsed and the impacts on environment and human health may be considerable. Certainly the use of alternative bleaching processes for paper manufacture is a positive development. However it is not clear whether the chemicals generated from the other processes in the manufacture, use, recycling and disposal of paper are as harmless as we are led to believe. Unfortunately, only time will tell.

References

1. Bellknap AM et al (2006) Identification of compounds associated with testosterone depressions in fish exposed to bleached Kraft pulp and paper mill chemical recovery condensates. Environ Toxicol Chem 25:2322–2333. http://www.ncbi.nlm.nih.gov/pubmed/16986786
2. Biedermann M, Grob K (2013) Assurance of safety of recycled paperboard for food packaging through comprehensive analysis of potential migrants is unreasistic. J Chromatog A 1293:107–119
3. Chiang G et al (2011) Health status of native fish (Percilia gillissi and Trichomycterus aerolatus) downstream of the discharge of effluent from a tertiary treated elemental chlorine-free pulp mill in Chile. Environ Toxicol Chem 30:1793–1809. http://www.ncbi.nlm.nih.gov/pubmed/21544863
4. Consonni D et al (2008) Mortality of a population exposed to dioxin after the Seveso, Italy accident in 1976: 25 years of follow up. Am J Epidemiol 16:847–858
5. Fukazawa M et al (2001) Identification and quantification of chlorinated bisphenol a in wastewater from wastepaper recycling plants. Chemosphere 44:973–997. http://www.ncbi.nlm.nih.gov/pubmed/11513431
6. Garg SK, Tripathi M (2011) Strategies for decolorization and detoxification of pulp and paper mill effluent. Rev Environ Contamin Toxicol 212:113–136. http://www.ncbi.nlm.nih.gov/pubmed/21432056
7. Haq I et al (2016) Evaluation of bioremediation of potentially lignolytic Serratia liquefaciens for detoxification of pulp and paper mill effluent. J Hazard Material 305:190–199
8. Hewitt LM et al (2006) A decade of research on the environmental impacts of pulp and paper mill effluents in Canada: sources and characteristics of bioactive substances. Toxicol Environ Health B CritRev 9:341–356
9. Hewitt LM et al (2008) Altered reproduction in fish exposed to pulp and paper mill effluents: roles of individual compounds and mill operating conditions. Environ Toxicol Chem 27:682–697
10. http://www.fda.gov/ForConsumers/ConsumerUpdates/ucm297954.htm
11. Humblet O et al (2008) Dioxins and cardiovascular mortality. Environ Health Perpsect 116:1443–1448
12. Jokinenen MP et al (2003) Increase in cardiovascular pathology in female Sprague-Dawley rats following chronic treatment with 2,3,7,9-tetrachlorodibenzo-p-dioxin and 3,3,4,4,5-pentachlorobiphenyl. Cardiovascul Toxicol 3:299–310
13. Lang IA et al (2008) Association of urinary bisphenol a concentration with medical disorders and laboratories in adults. JAMA 300:1303–1310

14. Myers JP et al (2009) Why public health agencies cannot depend on good laboratory practice for selecting data: the case of bisphenol a. Environ Health Perspect 117:309–315. http://www.ncbi.nlm.nih.gov/pubmed/19337501
15. National Institute for Occupational Safety and Health (NIOSH) review (2000). http://www.cdc.gov/niosh/docs/2001-107/pdfs/2001-107.pdf
16. Palanza P et al (2008) Effects of development exposure to bisphenol a on brain and behavior in mice. Environ Res 108:150–157
17. Pulp and Paper Resources and Information Site www.paperonweb.com/chemical.htm
18. Rappe C (1990) Environmentally stable chlorinated contaminants from the pulp and paper industry. IARC Sci Publ 104:341–335
19. Shibamoto T et al (2007) Dioxin formation from waste incineration. Rev Environ Comtam Toxicol 190:1–41. http://www.ncbi.nlm.nih.gov/pubmed/17432330
20. Takeuchi T et al (2004) Positive relationship between androgen and the endocrine disruptor, bisphenol a, in normal women and women with ovarian dysfunction. Endocr J 51:165–169
21. Terasaki M et al (2007) Occurrence and estrogenicity of phenolics in paper-recycling process water: pollutants originating from thermal paper in waste paper. Environ Toxicol Chem 26:2356–2366
22. Terasakin M et al (2008) Organic pollutants in paper-recycling process water discharge areas: first detection and emission in aquatic environment. Environ Pollut 151:53–59. http://www.ncbi.nlm.nih.gov/pubmed/17521789
23. Thacker NP et al (2007) Dioxin formation in pulp and paper mills of India 14, 225–226. Environ Sci Pollut Res Int 14:225–226
24. Uemura H et al (2008) Associations of environmental exposure to dioxins with prevalent diabetes among general inhabitants of Japan. Environ Res 108:63–66
25. Viel JF et al (2008) Risk of non-Hodgkin's lymphoma in the vicinity of municipal solid waste incinerators. Environ Health 7:51
26. Virk AP et al (2012) Use of laccase in pulp and paper industry. Biotechnol Prog 28:21–32
27. vom Saal FS, Hughes C (2005) An extensive new literature concerning low dose effects of bisphenol a shows the need for a new risk assessment. Environ Health Perspec 113:926–933
28. Walker NJ et al (2006) Comparison of chronic toxicity and carcinogenicity of 2,3,7,8-tetrachlorodibenzo-p-dioxins (TCDD) in 2-year bioassays in female Sprague-Dawley rats. Mol Nutr Food Fes 50:934–944
29. Wang X et al (2012) Formation and emission of PCDD/fs in Chinese non-wood pulp and paper mills. Environ Sci Technol 46:12234–12240. http://www.ncbi.nlm.nih.gov/pubmed/23067332
30. Yasuhara A et al (2005) A role of alkaline elements in formation of PCDDs, PCDFs, and coplanar PCBs during combustion of various paper samples. J Environ Sci Health A Tox Hazard Subst Environ Eng 40:991–1001. http://www.ncbi.nlm.nih.gov/pubmed?term=%22yasuhara%20a%22%20AND%20%22alkaline%20elements%22
31. Zhang Q et al (2004) Determination of polychlorinated dibenzo-p-dioxins and polychlorinated dibenzofurans in newsprints and copy papers by HRGC/HRMS. Se Pu 22:449–451. http://www.ncbi.nlm.nih.gov/pubmed?term=%22zhang%20q%22%20AND%20%22copy%20papers%22
32. Zhuang W et al (2003) Identification and confirmation of traces of chlorinated fatty acids in fish downstream of bleached Kraft pulp mills by gas chromatography with halogen specific detection. J Chromatog A 994:137–157

Chapter 9
Chemical Exposure in the Workplace

While we are all exposed, to varying degrees, to the vast number of chemicals generated by humankind, for most of us the level of exposure is not very great. However, there are certain situations where exposure for some sections of the community is much greater than the norm. A good example of this is in smokers who willingly accept the risks posed by exposure to the harmful chemicals generated by cigarettes because the habit makes us feel good. There are other situations where we also accept a greater degree of chemical exposure, not because it makes us feel good, but because we deem it necessary. The best example of this is the exposure that is a consequence of our working environment. Depending on the industry, occupational exposure to chemicals may be very significant. Moreover, there is little doubt that they can cause disease even though in some cases it may take years of court action before the matter is settled and the relationship between exposure and disease is accepted. The current view is that exposure to workplace chemicals does contribute to disease. In the case of the most feared disease, cancer, it has been suggested that as many as 5% of all cancers may be attributed to "work related carcinogens" (carcinogens are chemicals which in animals have been shown to cause cancer) [54, 55]. Of course this could be an underestimate because there is no way at present of confirming this (it could even be an overestimate).

There are many examples of this but perhaps the one that is most familiar to readers is the relationship between exposure to asbestos and mesothelioma, a rare type of lung cancer. The individuals who contracted the disease were those who were mostly exposed to asbestos through mining. More recently it has become apparent that exposure to asbestos also occurred in mining townships, shipyards, in workers in the automobile, aircraft and asbestos removal industries. Then there are other diseases such as asthma, which affects around 1 in 10 adults in developed countries, and skin diseases, such as contact and allergic dermatitis, in which there is clear evidence that workplace exposure to chemicals is an important factor in a significant proportion of sufferers. However, while these are the best known and well publicized examples of occupational exposure, they are really the proverbial tip of the iceberg as chemicals are used in many different industries and, even with the most stringent

© The Author(s), under exclusive license to Springer Nature Switzerland AG 2021
A. Poulos, *The Secret Life of Chemicals*,
https://doi.org/10.1007/978-3-030-80338-4_9

controls in place, some exposure occurs although the degree of exposure and the nature of the chemical or mixtures of chemicals varies according to the industry. What also varies is the period of exposure which depends on how long a worker remains in the industry.

The many reports on the impact of asbestos exposure on human health, and the ensuing court actions, have confirmed the potentially harmful effects of an unregulated work environment, a fact that was well known even during the industrial revolution in Victorian England. Despite the fact that the effects of the work environment on our health are known at some level they are perhaps not fully appreciated. There is now more than enough published information available in reputable international journals which confirms that asbestos is not the only chemical that we may be exposed to at work that is potentially harmful. Indeed, if the numbers of registered chemicals used throughout the world for various purposes runs in the tens of thousands, it is likely that workers in most industries have some degree of chemical exposure. But what are these chemicals, and is there really any evidence that, in the amounts present in the work environment, they can contribute to disease. In this chapter we will attempt to address some of these questions.

Workplaces and Chemical Exposure

Chemicals in the workplace come in different forms. They may be the complex mixtures that occur in dust particles – coal, asbestos, and silica dusts from mining, wood dust from woodworking, and agricultural fertilizer dusts are just a few of the many dusts we are exposed to in our working environment. Then there are liquids and fumes from various activities such as welding or liquids such as those used in degreasing and drycleaning. Exposure to the various chemicals occurs either through the lungs, absorption through the skin, from the skin into the mouth or via the nasal passgees directly into the brain. Whatever the method of exposure, occupational chemicals find their way into our bodies where they may contribute to the development of disease.

The diversity of chemicals in the workplace is truly astounding. While the variety of chemicals, and the degree of exposure of individual workers tends to be greater for workers in the manufacturing industries, there may still be considerable exposure even in workers whose activities strictly speaking do not involve manufacturing eg aircraft pilots, healthcare workers, doctors, publicans, policemen, armed services, firemen etc. Because of the great number of workplaces, it is proposed to focus on those in which there is evidence for a link between exposure and disease.

Occupation and Lung Diseases

The lungs are a major source of chemical exposure so it is therefore not surprising that the risk of diseases such as asthma and chronic obstructive lung disease (COPD) is increased in workers who are exposed to certain dusts and fumes [35, 60]. Occupationally induced asthma is thought to be the cause of around 15% of all those with asthma and COPD [9, 45]. There are indications as well that the conditions may be exacerbated by chemical exposure at work. Higher risk occupations for asthma include cleaners, nurses, printers, woodworkers, electrical processors, and workers in forestry and agriculture [39]. Chemicals which induce asthma in workers are many and varied. High-risk exposure to cleaning agents and pesticide exposure in developing countries appear to be as important as is exposure to isocyanates, cereal flour/grain dust, welding fumes, wood dust and, more recently, hairdressing chemicals, commonly reported in industrialised countries.

Mining

There is a considerable body of evidence for the impact of mining on lung disease in miners. The chemical exposure of workers in mining differs somewhat from workers in other industries such as manufacturing in that the agents are mostly natural products, for example coal, asbestos, iron ore etc. rather than the chemicals made by humankind. Nevertheless, even though they are in most instances natural products, they are chemical entities and, particularly in the form of dust particles and with continued exposure in over long periods of time, can induce the same sorts of processes in the human body that can lead to disease as non-natural chemicals. Moreover, while these dusts are natural products, they result from human activity.

Perhaps the type of mining known for its effects on the health of workers is coal mining. Coal mining has been known for many years to increase the risk of pneumoconiosis (meaning "dust in the lungs) or "black lung" [52]. This condition has been linked to the exposure to coal dust which builds up in the lungs over a period of time eventually triggering inflammatory reactions. If these reactions are severe enough they can result in the destruction of the lung tissue. US Government recognition of the role of coal dust in the development of black lung led to the enactment of the Federal Coal Mine Health and Safety Act in 1969 which provided for a greater emphasis on detection and prevention of the disease. The result was a very significant drop in the number of affected miners. However, it is worth pointing out that the disease has not been eradicted because in some parts of the US there has been a resurgence of the disease [8].

Another condition associated with mining is asbestosis which, as the name implies, is caused by a prolonged exposure to asbestos. Like coal dust, asbestos settles in the lungs and over time induces a chronic inflammatory response which in turn damages the sensitive lung tissue. The principal symptoms include shortness of breath and, in severely affected individuals, heart failure. Asbestos also increases the risk of lung cancer, and in particular mesothelioma. Court actions by sufferers against the mining companies have been upheld and have led to a general acceptance that there is a relationship between exposure to asbestos and mesothelioma. This has been confirmed even for other non-mining occupations, such as carpenters plumbers and pipe fitters, where there is direct exposure to asbestos or even household exposure through domestic activities such as handling or laundering of workers contaminated clothing [25, 41].

Silicosis is another lung disease believed to be caused by occupational exposure to silica dust. While the disease is associated with mining, it is also observed in workers in other workplaces including ceramic and pottery production, sandblasting, tunneling, glass manufacture, masonry, quarrying, and concrete and cement production [16]. The mechanism of action is similar to black lung and asbestosis ie traces of silica (which is the material found in sand and quartz) enter the lungs, and trigger a chronic inflammatory response which in turn damages the lung tissue causing gradual deterioration of function of the organ. Silicosis can also increase the risk of lung cancer [18]. The condition is not confined to exposure through mining but workers in many other occupations eg nonmetallic mineral manufacturing (stationary engineers, boiler operators), construction (brickmasons), and cut stone and stone product manufacturing (crushing, grinding and polishing machine setters) well known for their association with exposure to crystalline silica [44, 55]. Uranium mining is also believed to increase the risk of cancer, mostly of the lung [64]. Exposure to the radioactive gas radon appears to contribute to the increased risk although it is likely that there are other factors as well.

Occupation and Skin Diseases

Like the lungs, the skin is a major site of chemical exposure in the workplace [20]. Contact with chemicals may occur in different parts of the body although the hands are probably the most common. Workplace exposure to chemicals or natural products such as latex are a common cause of allergic contact dermatitis and involves damage to the skin by the immune system [31]. A related condition, irritant contact dermatitis, caused by damage to the skin but without the involvement of the immune system, is also common in the workplace [50].

Agriculture and Horticultural Workplaces

There are considerable difficulties in demonstrating unequivocally a link between an exposure to chemical agents and disease in farmers and workers in horticulture. The main reason for this is the large number of variables that can influence the findings. Thus, in the case of farmers, the variables include the age of the farmers, the type of farming, the type of herbicide, insecticide and fungicide used, the duration of the exposure, whether there are any so-called "confounding factors" ie factors like cigarette smoking, high blood pressure, excess alcohol consumption, and pre-existing diseases such as heart disease, diabetes, or cancer which could play a role in influencing the findings. There is also the so-called "healthy worker" effect which relates to the better general health and fitness of farmers because of their more active lifestyle. The studies reported rely on comparing the health of a number of farmers with a corresponding group of non-farmers who are matched as closely as possible in age, sex, general health, alcohol consumption, smoking etc. and then, using statistics to determine whether the any differences observed are "significant" or not. The word "significant" in a statistical sense really means the likelihood that any differences found have arisen purely by chance rather than as a result of a difference in the two groups caused by some factor or factors in the working environment ie pesticide exposure.

There have been numerous studies carried out over the years on the health of farmers and, as expected, because of the great number of variables, the conclusions are not unequivocal. Just to illustrate the point – the word "farmer" can mean someone growing all sorts of fruits, grains, vegetables, mushrooms, or someone rearing sheep, cattle, pigs, poultry, and even fish. Depending on the type of farming, the farmer may be exposed to various synthetic chemicals such as insecticides, herbicides, or fungicides, or they may be exposed to dusts from composts, grain, hay, soil, manure etc. Despite this great variability, there are significant numbers of reports that consistently point to an increased risk of abnormalities in lung function, skin conditions, certain cancers, and Parkinson's disease in farmers [19, 21, 26, 51]. When one then includes the results of experiments which show the toxic effects of many of the pesticides in different animal species, and others demonstrating that many of the pesticides have killed or made people very ill, it becomes difficult to argue that the only studies that are valid are those that show no relationship between pesticide exposure and disease. Moreover, even the governments around the world which have consistently supported the use of pesticides have, at times, withdrawn the registration of certain pesticides indicating that they believed that there were sufficient concerns to warrant discontinuation of use [12].

Woodworking

The processing of wood, such as sanding and sawing, can generate particles which may vary in shape and also in size, from fine to relatively thick. Workers are exposed to the particles which contain a variety of different chemical substances naturally present in wood, in addition to other chemicals that may have been used for treatment of the wood. Parts of the body that are most exposed by woodworkers include the skin, nasal passages and sinus, and the lungs. Apart from asthma mentioned earlier, there is evidence that wood workers have higher than normal risk of developing nasal cancers indicating that some of the components of wood may be carcinogenic [2]. Indeed, the International Agency of Research into Cancer (IARC) has listed wood dust as a carcinogen [67].

Food Handling

Baker's asthma is one of the commonest forms of occupational induced asthma. It is caused by exposure to flour dust and is not confined to bakers but includes anyone exposed to flour including confectioners, pastry factory workers, millers, and cereal handlers. Upon inhalation of the dust, some of the proteins in flour trigger an allergic response in the lung and this can lead to asthma. Wheat flour is the main cause of baker's asthma although other flours, such as rye, oats, and barley, can also cause an allergic response [34, 56].

Industrial Activities

The number of workplaces which would fit into the definition of "industrial activities" is very large and it is therefore not possible to cover each within this chapter. It is perhaps best to focus on a few chemical pollutants that are known to increase the risk of disease and to indicate those industries in which workers are exposed to these particular pollutants. Because of the vast number of chemicals that are used in industry, some of which show some evidence, at least in animals, of having the potential to harm, it will not be possible to list all. Instead, it is proposed to identify some of the chemicals known to increase the risk of disease in the workplace, and to indicate the workplaces where it could be expected that they would be encountered.

Exposure to Metals and Disease

The three metals for which there is some evidence of a link between occupational exposure and disease are cadmium, arsenic, and beryllium.

Cadmium

Most cadmium is used for the manufacture of batteries. It also used in pigments, plastics, solders, electroplating and for semiconductors. It is a contaminant in rock phosphate, the raw material used to make many phosphate containing fertilizers and this use has led to some contamination of soils, particularly after repeated applications of fertilizers. It is known carcinogen is animals and in high doses has been shown to induce different cancers, including lung, liver, and adrenal cancers [48]. Cadmium is toxic in high doses to humans, with reports of effects on nerve function, bone and kidney, and lung when workers are exposed to cadmium fumes in the silver jewellery production [53, 59]. With chronic exposure ie exposure to smaller amounts over longer periods of time, there are indications of abnormalities in kidney function as determined by changes in the urine and blood levels of certain proteins (beta-2 microglobulin) [36]. However, more recently, it has been suggested that the apparent changes that have been observed from chronic exposure to very small amounts may just simply be a normal response of the kidney [1]. There are also indications that chronic exposure can increase the risk of bladder cancer [37].

Arsenic

Arsenic has been discussed in an earlier chapter of this book. To reiterate, it is a metal which occurs naturally in small amounts throughout the Earth's crust. It can leach from rocks into water causing the contamination of drinking water in a number of different countries including Bangladesh, India, and China. It had been used extensively in pesticides in the past although its use in agriculture is now limited since the advent of synthetic pesticides. However, it is used in many different industries, including the production of alloys, paints, glass, batteries, pesticides, and semiconductors. It has even been used as an additive to chicken and pig feeds. It is also generated is significant amounts in lead and copper smelting. It was also formerly used extensively as a wood preservative and as a pesticide in agriculture but these uses are now limited. Exposure to arsenic in the workplace is therefore at least theoretically possible in any of these activities.

Arsenic is a well known poison producing nausea, vomiting, abdominal pain, and diarrhea. We know something of the effects of chronic arsenic exposure ie exposure to small amounts over long periods of time, because of studies that have been carried out on people who live in those parts of the world where there are high levels of arsenic in drinking water. Increased pigmentation and thickening of the skin, affects on lung and liver function, anaemia, and an increased risk of diabetes, and skin, lung and bladder cancers are associated with chronic exposure to arsenic [27, 49, 63]. It could be expected that workers in some of the industries referred to above would be exposed to significant levels of arsenic and this would lead to an increased risk of cancer and skin abnormalities. Certainly, workers involved in the manufacture of arsenic containing pesticides, mining, and smelting have been reported to have the same sorts of skin abnormalities, ie the changes in pigmentation and the thickening, and effects on nerve function, which have been observed in individuals exposed to arsenic contaminated drinking water [28, 61]. It would indeed be surprising if workers in other industries did not have an increased risk of disease.

Beryllium

Beryllium is a metal with a variety of industrial uses including the production of alloys, gyroscopes, computer equipment and watch springs, as well as in X-ray detection diagnostics, nuclear reactors, and X-ray lithography. The numbers of workers exposed to beryllium are not insignificant with an estimated number of more than a hundred thousand exposed in the US [30]. Chronic exposure to beryllium-containing dusts can result in the sensitization of delicate lung tissue eventually leading to inflammation and eventual damage to lung tissue with the development of fibrosis (replacement of lung tissue with connective tissue) thereby affecting the function of the organ. Not all workers exposed to beryllium develop lung disease and there are indications that genetic factors can contribute to the disease process [23, 57].

Indium

Indium is a rare metal used in the production of alloys, in the electronics industries, and in medical imaging. Workplace exposure to indium increases the risk of emphysematous progression of the lungs, a condition characterized by the gradual accumulation of air pockets in the organ making it progressively difficult to breathe [46].

Chromium

Chromium is a metal that is used in leather tanning, chrome plating, and stainless steel welding industries. It exists in two forms – the so-called "trivalent" and "hexavalent" forms. The latter is considered to be toxic. There are indications that occupational exposure to the hexavalent form may increase the risk of different cancers, in particular lung and stomach cancers [66].

Non-metallic Exposure

Formaldehyde

Formaldehyde is a very simple substance, unique in that it contains only one carbon, one oxygen and one hydrogen atom. It is a highly reactive substance with the capacity to interact with a variety of the chemical substances that make up the human body, in particular proteins. It is this particular property that forms the basis for its use as an embalming fluid and preservative for cadavers. It is used in the manufacture of resins for adhesives in plywood and carpets, the manufacture of certain industrial chemicals, paper, textiles, paints, furniture and insulation. According to the International Agency for Research on Cancer (IARC) established by the World Health Organisation, formaldehyde is considered to carcinogenic ie it has the ability to cause cancer in humans and animals. More than two million people in the USA are occupationally exposed to formaldehyde so the workplace exposure is considerable. There have been suggestions that there is an increased risk of nasopharyngeal (the area the back of the nose leading into the throat) cancer but this has not been confirmed by more recent studies [5, 29, 47]. However, the evidence for a link between formaldehyde exposure and leukemia appears more persuasive [17, 58]. In addition to its possible carcinogenic effects, there are indications that formaldehyde exposure can increase the risk of asthmatic attacks in asthmatics whose asthma is triggered by certain environmental agents such as dust mites although, again, this has been disputed [14, 68].

Benzene

Benzene is an aromatic hydrocarbon that occurs naturally in crude oil. The term "aromatic hydrocarbon" indicates that the carbon atoms that make up its chemical structure are joined together to form a six sided (hexagon) ring. Benzene exposure is widespread and employees are exposed in a number of industries including petroleum refineries, chemical and plastics plants, paint, rubber, tyre, asphalt manufacture and shoe manufacture, and the printing and steel industries. IARC considers that benzene is a human carcinogen.

There is considerable evidence linking benzene exposure with cancer. In particular, the evidence for an association between leukemia (a cancer of the white cells in the blood) and benzene exposure is highly persuasive [38]. There are also indications of a link between benzene exposure and non-Hodgkin lymphoma (also a cancer of white cells) and kidney cancer but the evidence for this is no so clear cut [11, 62]. The risk of developing leukemia appears to be related to the degree and the duration of exposure.

Diacetyl

The supermarket shelves are full of products that contain what are loosely called "flavorings". These are substances that are mostly either extracted from foods such as fruits or dairy, or are chemically synthesized, and which, when added to food produce a taste that is reminiscent of the food itself. This is because the flavor of individual foods such as, to take a simple example, raspberries, is determined by the presence of chemical substances which stimulate the senses of smell and taste in a way that is unique for raspberries and not, for example, for oranges or apples. Of course there are other factors, such as texture, that inform the brain that we are eating raspberries and not apples. Because certain flavorings are so desirable, it is not surprising that food manufactures have resorted to increasing the appeal of foods to the consumer by adding flavor chemicals. After all, why add raspberries, which are seasonal, relatively expensive, and with a short shelf life, to something when you can mimic the flavor by adding a simple chemical or chemical mixture?

Another unique flavor that is much used by the food industry is butter flavor. And there may be quite genuine reasons for adding butter flavor to a food rather than butter because of the perception that the saturated fats in butter may increase the risk of heart disease. One of the chemicals that is responsible for the taste of butter is diacetyl, a very simple chemical substance that is easy to produce. It is often added to margarines to provide a more acceptable flavor for consumers. Unfortunately, even though diacetyl is apparently "natural", it does not mean that it is completely safe. Experiments with animals have demonstrated that it causes quite severe inflammation of the nasal passages, the trachea (the tube that goes down the throat into the lungs), and the lungs [33]. In an inhaled form, quite small doses (200–300 parts per million) can produce a significant effect. It does not just affect animals because there is very good evidence that workers that are involved in producing diacetyl as a food flavoring have a diminished lung function with increased coughing, shortness of breath, and asthma [65]. Exposure can also cause blockage of the bronchioles (the airways) in the lung and eventually lead to the serious lung condition, bronchiolitis obliterans. This condition has also been reported in workers in the popcorn industry, where butter flavoring is often added to the corn, hence the common name for the condition "popcorn maker's lung" [40]. What is worrying is that it is not just popcorn makers can develop the condition, but consumers may also be affected [22].

Diesel Exhaust

Petrol and diesel are the two main fuels used to to drive internal combustion engines in cars and trucks. They are both derived from oil and consist mainly of hydrocarbons, but there are significant differences in their chemical composition and also in the composition of the gases formed by combustion and subsequently released into the environment as exhausts. While petrol is used mostly in domestic vehicles, diesel is mostly used in mining, trucking and transport industries [4]. It is believed that diesed exhausts produce up to seven times the amounts of secondary aerosols (mentioned in Chap. 6), particles or droplets formed by the interaction of the components in the exhausts gases with substances present in air [24]. The chemical composition of these aerosols is highly complex.

Other Chemicals

A few more examples of occupational chemicals known to increase the risk of disease are shown below.

Diisothiocyanates are a group of chemicals that are used for surface coatings, polyurethane foams, adhesives, resins, automotive parts and as curing agents in foundries. Exposure can occur either through inhalation or through skin contact. They are known to cause asthma in some people and possibly chronic obstructive lung disease, a severe inflammatory lung condition [6, 7, 43].

Glutaraldehyde has been used extensively in the health care industry where it is used as a disinfectant for dialysis and surgical instruments, suction bottles, bronchoscopes and ear nose and throat instruments. It is also used in pathology laboratories. It can cause headaches, burning eyes, throat irritation, dermatitis and asthma [3, 15].

Aromatic amines include benzidine, naphthylamine, and 4-aminobiphenyl and are products of the dye and chemical industries. Many of these substances are considered to be carcinogenic in animals [42]. There have been a number of reports showing that chronic occupational exposure of humans over long periods of time can increase the risk of bladder cancer [13].

Vinyl chloride is the chemical precursor (a monomer) of PVC (or polyvinyl chloride), a common plastic used to make a variety of products including packaging materials, automotive parts, construction materials and furniture.Workers in the petrochemical and plastics industries in particular are exposed.to the chemical. Vinyl chloride is a known carcinogen is animals and there is good evidence that chronic exposure can increase the risk of a rare type of liver cancer in humans [10].

What Does All This Mean?

It has been known for a long time that the work environment can be hazardous. In the case of mining, there is always the risk of death due to an explosion or mine collapse, although the industy is now better regulated and the risks are not as great as they used to be, certainly in the developed world. Much more common are the back ailments, repetition strain injury, accidents, stress, and loss of hearing that can develop at work. However, while most people know that physical injuries can result from what they do at work, they do not fully appreciate the risks of exposure to workplace chemicals. The main reason for this is that, unlike the physical injuries sustained at work, chemical exposure can take a decade or more to develop into disease. Moreover, during this time, there may be no indication of any disease. In a way, it is like the risks posed by smoking which can take decades to develop into disease.

The very good example of the long lead times for chemical-related disease is the mesothelioma that develops from asbestos exposure and which can take up to three or four decades to develop. While, through better regulatory procedures, it is possible to reduce the risk of disease caused by acute chemical exposure, it is much more difficult to regulate against long term chronic exposure, because for most occupational chemicals little is known of the period of exposure, or the amounts of the chemical actually required to cause disease. One way this is being handled by governments is by reduction, through regulation, of the degree of chemical exposure, particularly for chemicals known to have, for example, carcinogenic properties. This can certainly help but no government on earth is able predict what is likely to happen after 10, 20 or 30 years of continuing exposure to amounts that are deemed to be insignificant. It is exceedingly difficult to design animal experiments that can predict what is likely to happen in humans after a decade or more of exposure, and, moreover, humans are not rats or mice, the animals most frequently used for long term studies. Epidemiological studies, involving a comparison of the health of workers in different industries, are useful and can point to problems in a particular industry, but the chemical agent or agents responsible for increasing the risk of disease are not always known. And, due to the large number of variables which have been discussed throughout this book, there is considerable inconsistency in these types of studies so any degree of certainty regarding the impact of a particular chemical on a worker's health is most unlikely.

Despite this, there are ways of reducing the risk of workplace chemical disease. There are clearly workplaces, and some of these have been identified above, for example mining and certain industries, in which exposure to specific chemicals or dusts have been shown unequivocally to increase the risk of disease, particularly cancer and lung diseases. Therefore employees in these industries should try, wherever possible, to determine whether the employer has in place appropriate occupational health and safety protocols to protect them. In other industries or workplaces, where employees are aware that there is chemical exposure, reference to lists of chemicals which have been published by reputable organizations such as the IRAC would certainly help employees and their employers to determine whether they are exposed to substances known to be carcinogens. While it may not always be practical, if a worker is aware that he or she is exposed to one of these, a decision can be made as to whether it is worth the risk to work in this environment.

References

1. Akerstrom M et al (2012) Associations between urinary excretion of cadmium and proteins in a nonsmoking population: renal toxicity or normal physiology. Environ Health Perspect. http://www.ncbi.nlm.nih.gov/pubmed/23128055
2. Alonso-Sardon M et al (2015) Association between occupational exposures to wood dust and cancer: a systematic review and meta-analysis. PLoS One 10:e0133024
3. Arrandale VH et al (2012) Occupational contact allergens: are they also associated with occupational asthma. Am J Ind Med 55:353–360. http://www.ncbi.nlm.nih.gov/pubmed/22238032
4. Attfield MD et al (2012) The diesel exhausts in miners study: a cohort mortality study with emphasis on lung cancer. J Natl Cancer Inst 104:869–883

5. Bachand AM et al (2010) Epidemiological studies of formaldehyde exposure and risk of leukemia and nasopharyngeal cancer: a meta-analysis. Crit Rev Toxicol 40:85–100. http://www.ncbi.nlm.nih.gov/pubmed/20085478

6. Baur X et al (2012) Bronchial asthma and COPD due to irritants in the workplace- an evidence-based approach. J Occup Med Toxicol 7:19. http://www.occup-med.com/content/pdf/1745-6673-7-19.pdf

7. Bello D et al (2007) Skin exposure to isocyanates: reasons for concern. Environ Health Perspect 115:328–335. http://www.ncbi.nlm.nih.gov/pubmed/17431479

8. Blackley DJ et al (2016) Resurgence of Progressive Massive Fibrosis in Coal Miners – Eastern Kentucky. MMWR Morb Mortal Wkly Rep 65, 1385–9

9. Blanc PD (2012) Occupation and COPD. J Asthma 49:2–4

10. Brandt-Rauf PW et al (2012) Plastics and carcinogenesis: the example of vinyl chloride. J Carcinog 11:5. http://www.ncbi.nlm.nih.gov/pmc/articles/PMC3327051/

11. Brautbar N et al (2006) Occupational kidney cancer: exposure to industrial solvents. Ann N Y Acad Sci 1076:753–764

12. Browner C (2000, June) Dursban announcement. http://www.epa.gov/aboutepa/dursban-announcement

13. Burger M et al (2012) Epidemiology and risk factors of urothelial bladder cancer. Eur Urol. http://www.ncbi.nlm.nih.gov/pubmed/22877502

14. Casset A et al (2006) Inhaled formaldehyde exposure: effect on the bronchial response to mite allergen in sensitized asthma patients. Allergy 61:1344–1350. http://www.ncbi.nlm.nih.gov/pubmed/17002712

15. Centers for Disease Control and Prevention (2001) Glutaraldehyde – occupational hazards in hospitals. NIOSH Publications and Products. http://www.cdc.gov/niosh/docs/2001-115/

16. Centres for Disease Control and Prevention (CDC) (2005) Silicosis mortality, prevention and control – United States 1968–2002

17. Checkoway H et al (2012) Critical review of the epidemiologic evidence on formaldehyde exposure and risk of leukemia and other lymphohematopoietic malignancies. Cancer Causes Control 23:1747–1766. http://www.ncbi.nlm.nih.gov/pubmed/22983399

18. de Klerk NH, Musk AW (1998) Silica, compensated silicosis, and lung cancer in Western goldminers. Occup Environ Med 55:243–248

19. Dhillon AS et al (2008) Pesticide/environmental exposures and Parkinson's disease in East Texas. J Agromedicine 13:37–48

20. Diepgen TL, Kanverva L (2006) Occupational skin diseases. Eur J Dermatol 16:324–330

21. Dreither J, Kordysh E (2006) Non-Hodgkin lymphoma and pesticide exposure: 25 years of research. Acta Haemotol 116:153–164

22. Egilman DS, Schilling JH (2012) Bronchiolites obliterans and consumer exposure to butter flavored microwave popcorn: a case series. Int J Occup Environ Health 18:29–42. http://www.ncbi.nlm.nih.gov/pubmed/22550695

23. Fontenot AP, Maier LA (2005) Genetic susceptibility and immune mediated destruction in beryllium-induced disease. Trends Immunol 26:543–549

24. Gentner DR et al (2012) Elucidating secondary organic aerosol from diesel and gasoline vehicles through detailed characterization of organic emissions. Proc Natl Acad Sci U S A 109:18318–18323

25. Goswamin E et al (2013) Domestic asbestos exposure: a review of epidemiological and exposure data. Int J Environ Res Public Health 10:5629–5670

26. Greskevitch M et al (2007) Respiratory diseases in agricultural workers: mortality and morbidity statistics. J Agromedicine 12:5–10

27. Guha Mazumder DN (2008) Chronic arsenic toxicity and human health. Indian J Med Res 128:436–447. http://www.ncbi.nlm.nih.gov/pubmed/19106439

28. Hamada T, Horiguchi S (1976) Occupational chronic arsenical poisoning. On the cutaneous manifestations. Sangyo Igaku 18:103–115. http://www.ncbi.nlm.nih.gov/pubmed/1035665

29. Hauptmann M et al (2004) Mortality from solid cancers among workers in formaldehyde industries. Am J Epidemiol 159:1117–1130. http://www.ncbi.nlm.nih.gov/pubmed/20798648

30. Henneberger PK et al (2004) Industries in the United States with airborne beryllium exposure and estimates of the number of current workers potentially exposed. J Occup Environ Hyg 1:648–659

31. Holness DL (2014) Occupational skin allergies. Testing and treatment (the case of occupational allergic contact dermatitis). Curr Allergy Asthma Rep 14:410

32. http://www-ncbi-nlm-nih-gov.proxy.library.adelaide.edu.au/pubmed/24946105

33. Hubbs AF et al (2012) Respiratory and olfactory cytoxicity of inhaled 2, 3-pentanedione. Am J Pathol 181:829–844. http://www.ncbi.nlm.nih.gov/pubmed/22894831

34. Hur GY, Park HS (2015) Biochemical and genetic markers in occupational asthma. Curr Allergy Asthma Rep 15:488

35. Jeebhay MF, Quirce S (2007) Occupational asthma in the developing and industrialised world: a review. Int J Tuberc Lung Dis 11:122–133

36. Kawasaki T et al (2004) Markers of cadmium exposures in workers in a cadmium pigment factory after changes in their exposure conditions. Toxicol Ind Health 20:51–55

37. Kellen E et al (2007) Blood cadmium may be associated with bladder carcinogenesis: the Belgian case control study on bladder cancer. Cancer Detect Prev 31:77–82

38. Khalade A et al (2010) Exposure to benzene at work and the risk of leukemia: a systematic review and meta-analysi. Environ Health 28:9–31. http://www.ncbi.nlm.nih.gov/pubmed/20584305

39. Kogevinas M et al (2007) Exposure to substances in the workplace and new-onset asthma: an international prospective population-based study (ECRHS-II). Lancet 370:336–341

40. Kreiss K et al (2002) Clinical bronchiolitis obliterans in workers in a microwave popcorn plant. N Engl J Med 347:330–338

41. Lehman EJ et al (2008) Proportional mortality study of the united Association of Journeymen and Apprentices of the plumbing and pipe fitting industry. Am J Ind Med 51:950–963

42. Letasiova S et al (2012) Bladder cancer, a review of the environmental risk factors. Environ Health 11(Suppl 1):S11. http://www.ncbi.nlm.nih.gov/pubmed/22759493

43. Mapp CE et al (2005) Occupational asthma. Am J Resp Crit Care Med 172:280–305. http://ajrccm.atsjournals.org/content/172/3/280.full

44. Mazurek JM et al (2015) Notes from the Field:Update: Silicosis Mortality. United States, 1999–2013. MMWR Morb Mortal Wkly Rep 64, 653–4

45. Mazurek JM et al (2015) Work-related-asthma – 22 states, 2012. MWR Morb Mortal Wkly Rep. 64, 343–6

46. Nakano M et al (2014) Five year cohort study: emphysematous progression in indium-exposed workers. Chest 146:1166–1175

47. Final report on carcinogens background document for formaldehyde. National Toxicology Program. Rep Carcinoge Backgr Doc 2010 Jan 10–5981 http://www.ncbi.nlm.nih.gov/pubmed/20737003

48. National Toxicology Program, Department of Health and Human Services. Report on Carcinogens Twelfth Edition "Cadmium and Cadmium Compounds" http://ntp.niehs.nih.gov/ntp/roc/twelfth/profiles/Cadmium.pdf

49. Navas-Acien A et al (2008) Arsenic exposure and prevalence of type 2 diabetes in US adults. JAMA 300:814–822

50. Nicholson PJ (2011) Occupational contact dermatitis: known knowns and known unknowns. Clin Dermatol 29:325–330

51. Nordgren TM, Bailey KL (2016) Pulmonary health effects of medicine. Curr Opin Pulm Med

52. O'Brien C, McKillop C (2017) Black lung detection in Queensland "deliberately underfunded, under-resourced. http://www.abc.net.au/news/2017-03-22/black-lung-interim-report-handed-down-in-queensland-parliament/8376076

53. Panchal L, Valdeeswar P (2006) Acute lung injury due to cadmium inhalation- a case report. Indian J Pathol Microbiol 49:265–266

54. Pukkala E et al (2012) Occupation and cancer- follow-up of 15 million people in five Nordic countries. Acta Oncol 48:646

55. Rushton L et al (2008) The burden of cancer at work: estimation as the first step to prevention. Occup Environ Med 65:789–800

56. Salcedo G et al (2011) Wheat allergens associated with baker's asthma. J Investig Allergol Clin Immunol 21:81–92. Allergy Asthma Immunol Res. 2013 Nov; 5(6):348–56. https://doi.org/10.4168/aair.2013.5.6.348. Epub 2013 Jun 25. http://www.jiaci.org/issues/vol21issue2/1.pdf

57. Schuler CR et al (2012) Sensitization and chronic beryllium disease at a primary manufacturing facility, part 3: exposure-response among short term workers. Scand J Work Environ Health 38:270–281. http://www.ncbi.nlm.nih.gov/pubmed/21877099

58. Schwilk E et al (2010) Formaldehyde and leukemia: an updated meta-analysis of bias. J Occup Environ Med 52:878–896

59. Sethi PK, Khandelwal D (2006) Cadmium exposure: health hazards of silver cottage industry in developing countries. J Med Toxicol 2:14–15

60. Silverman DT et al (2012) The diesel exhaust in miners study: a nested case-control study of lung cancer and diesel exhaust. J Natl Cancer Inst 104:855–868

61. Sinczuk-Walczak H et al (2010) Effects of occupational exposure to arsenic on the nervous system: clinical and neurophysiological studies. Int J Occup Med Environ Health 23:347–355. http://www.ncbi.nlm.nih.gov/pubmed/21306980

62. Smith MT et al (2007) Benzene exposure and risk of non-Hodgkin lumphoma. Cancer Epidemiol Biomark Prev 16:385–391

63. Taeger D et al (2009) Major histopathological patterns of lung cancer related arsenic exposure in German uranium miners. Int Arch Occup Environ Health 82:867–875. http://www.ncbi.nlm.nih.gov/pubmed/19020892

64. Tomasek L et al (2008) Lung cancer in French and Czech uranium miners: radon-associated risk of low exposure rates and modifying effects of time since exposure and age of exposure. Radiat Res 169:125–137

65. van Rooy FG et al (2009) A cross sectional study of lung function and respiratory symptoms among chemical workers producing diacetyl for food flavourings. Occup Environ Med 66:105–110

66. Welling R et al (2015) Chromium VI and stomach cancer: a meta-analyis of the current epidemiological evidence. Occup Environ Med 72:151–159

67. WHO International Agency for Research on Cancer Monograph on the Evaluation of Carcinogenic Risk to Humans. Wood Dust and Formaldehyde Vol 62, 1995

68. Wolkoff P, Nielsen GD (2010) Non-cancer effects of formaldehyde and relevance for setting an indoor air guideline. Environ Int 36:788–799. http://www.ncbi.nlm.nih.gov/pubmed/20557934

Chapter 10
Fluorocarbons

The Collins English dictionary describes fluorine as "a poisonous, strong-smelling pale yellow gas that is the most reactive of all elements". There are similarities with chlorine which is also a gas (greenish) and highly reactive. Whereas the element chlorine is present in large amounts on Earth, mainly combined with the element sodium as common salt, fluorine has a more limited occurrence. The principal sources of fluorine include fluorite, a mineral made up of two atoms of fluorine and one atom of calcium, and fluorapatite. On the other hand, fluorocarbons are substances mostly made by humankind and, as the name implies, contain mixtures of carbon and fluorine. Some fluorocarbons also contain other elements such as oxygen, hydrogen and chlorine. There is a further division because fluorocarbons are divided into a further two distinct categories – the so called "perfluorocarbons" and the fluorocarbons. Most of the carbon atoms in all living things are attached to one or more hydrogen atoms, whereas the perfluorocarbon carbon atoms mostly contain attached fluorine rather than hydrogen atoms. This difference in chemical structure has important implications, both in relation to their properties as well as their chemical stability.

From a scientist's point of view, what makes fluorine unique is the small size of the atom and its very high reactivity. On the other hand, and this is one reason why fluorocarbons are so useful industrially, the chemical bond that is formed when an atom of fluorine combines with an atom of carbon is very stable which means that it is fairly inert to attack by chemicals. One of the principal industrial uses of fluorocarbons is as refrigerant gases, for air conditioning, and as propellants. In combination with carbon and chlorine it forms chlorofluorocarbons, refrigerants that have been implicated in their effects on the ozone layer. As a consequence, the earlier chlorofluorocarbons have been largely replaced by other fluorocarbons. There is yet another, and surprising use for fluorocarbons -two of the better known drugs, the antidepressant Prozac, and the cholesterol lowering Lipitor, are both fluorocarbons.

Industrially, fluorocarbons are particularly useful because they can form polymers. We have discussed polymers in the chapter on plastics and how they are formed from simpler substances termed "monomers". Fluorocarbons can also act as monomers to form a variety of polymers, the best known being Teflon or to use the chemical term "tetrafluoroethylene". There may be a thousand or more monomers in a single molecule of a fluorocarbon polymer like Teflon. As discussed in an earlier chapter, if the number of monomers is small (eg 10 or less) the substance is referred to as a "oligomer". For fluorocarbons, the corresponding term is a "fluorocarbon telomer". There are potentially many different telomers, the best known examples include perfluorooctanoic acid (PFOA) and pertfluorooctane sulphonate (PFOS). Because of their resistance to high temperatures, chemical reagents, stains and water, fluorocarbon polymers and telomers have a variety of uses. These include insulation, the manufacture of non-stick cookware, beauty products such as nail polish and curling irons, automobile components, carpets, paper coatings, electrical and electronic equipment, building materials etc. are just some of the uses of Teflon and related products [53]. Because of its detergent-like properties, PFOA is used in the manufacture of Teflon and other fluorocarbon polymers. PFOS has been used as a fabric protector, in fire fighting foams, hydraulic fluids, adhesives, waxes, and paper.

Release of Fluorocarbons into the Environment

We have already discussed in some detail how many chemicals used for a variety of industrial and other activities eventually find their way into the environment. Whether they are insecticides, PCBs, PBDEs, heavy metals such as lead and mercury, plastics and their additives, motor vehicle exhausts, or chemicals used in paper manufacture, the result is the same.- they end up in the air, in the soil, and in rivers and seas, including some of the most remote places on the planet and far removed from their source. Fluorocarbons are no different because they are really found everywhere we look [1–5]. Like PCBs and PBDEs, they are normally present as mixtures of different fluorocarbons, and PFOA and PFOS also have considerable environmental stability [54, 55]. It is a little sobering to note that, even our homes are not free of fluorocarbon contaminants because analysis of vacuum cleaner dust from homes in different parts of the world, has revealed as many as 16 different fluorocarbons [3].

The release of fluorocarbons into the environment, and in particular PFOA, was the subject of a legal class action between Jack W Leach et al. and Dupont de Nemours and Co, in 2002 [12]. The action related to the contamination of drinking water in six water districts in two states in the USA through the release of fluorocarbons, mainly PFOA, from a DuPont facility near Parkersburg, West Virginia. Analysis of the water showed that the levels of fluorocarbons were at least 6–8 fold greater in the drinking water than anywhere else in the USA. A pre-trial settlement was reached providing $70 million for health and education for class members who included residents who drank the tap water from one, or more than one, of the six districts. The settlement also included a filtering system to remove PFOA as well as the medical monitoring of residents exposed to the contaminated water.

Are Fluorocarbons Found in Food and Water?

There are numerous studies showing that fluorocarbons are present in very small amounts in the food of people from many different countries [7–9]. Indeed, it would seem that fluorocarbons are now ubiquitous. Many different foods contain tiny amounts, mostly of mixtures of different fluorocarbons and, again, PFOS and PFOA appear to be two of the most important fluorocarbon contaminants. Analysis of the different food components has demonstrated that they are present in many different foods although fish in particular appears to be an important source [37]. One estimate for the combined total dietary intake of PFOA and PFOS is less than 0.1 of a millionth of a gram [6]. Obviously, as the level of these fluorocarbons varies greatly in the different foods, this figure is very much dependent on an individual's diet. These figures also do not take into account the presence of fluorocarbons other than PFOS and PFOA.

It is not entirely clear how fluorocarbons contaminate food but food packaging materials, especially paper and paperboard containers used for oily fast foods, bags for microwaving popcorn, and non stick saucepans are important sources [40, 41]. Because of the migration of fluorocarbons used in packaging into food and their potentially harmful effects, pressure from government agencies such as the US EPA and FDA has led to the gradual withdrawal of some but the use of the so-called "shorter chain" fluorocarbons ie fluorocarbons with fewer fluorine atoms, continues.

Mention has already been made in relation to the contamination of drinking water in certain districts of the US. The contamination of tap water, and beverages made from tap water, with fluorocarbons appears to be ubiquitous [10, 11, 36]. Fluorocarbons have been found in spring water and bottled water although the amounts are lower than tap water [11, 21, 38]. The total amount of combined PFOS and PFOA taken up per day through water consumption has been estimated to be around 0.2 millionths of a gram per day which is not very much although this amount is taken every day – day in and day out throughout the year [16].

Are Fluorocarbons Present in Human Tissues and Body Fluids?

The answer to this question is most definitely yes. Mixtures of fluorocarbons are found- again PFOS and PFOA are major components- in plasma (ie the liquid component of blood) and in various human tissues [13, 14, 15, 39]. Assuming the average amount of plasma in the human body is about 2.5 litres, the combined amounts of PFOS and PFOA are quite small (less than 70 millionths of a gram) and have decreased in the last decade. However, it is important to bear in mind that these figures are hardly representative because they were obtained from very few individuals and, because of differences in fluorocarbon exposure, some people may have less and others may have a lot more.

How Do Fluorocarbons Get into Our Bodies?

Because fluorocarbons, or at least the fluorocarbons found in the human body and body fluids are not normal components, we have to conclude that they have been taken up, either via the diet, the mouth (other than diet), through inhalation, or the skin. There is no question that each of us will take in some fluorocarbons from the food that we eat, but what about inhalation and skin exposure? As far as inhalation (ie breathing in contaminated air), there is no doubt that fluorocarbons are present in both the air in our homes as well as outdoor air. A comparison of the fluorocarbons in air and food has demonstrated that there are differences in the types of fluorocarbons in these two sources which is not surprising because it could be expected that the more volatile fluorocarbons would predominate in air. In fact this is what has been found [17]. Also, there is evidence that the levels of these substances are higher in indoor than outdoor air, particularly in shops selling certain consumer products eg furniture, carpets, and outdoor equipment [18].

Treatment of household carpets with stain repellant materials containing fluoro-carbons (Scotchguard) was reported to greatly increase blood levels of these sub-stances in all family members of a Canadian household, presumably because of their increased exposure [19]. The main fluorocarbon detected in the blood was PFHxS, normally a minor component, but known to be present in significant amounts in Scotchguard formulations. It was speculated that most of the fluorocarbons in the blood were either inhaled or ingested as a dust. Household dusts have been shown repeatedly to contain fluorocarbons and this has led to speculation that the larger amounts of fluorocarbons in the blood of some children is at least in part due to the ingestion of dust and the hand to mouth transfer from treated carpets [20].

It is not known how much of the fluorocarbons in our bodies are absorbed through the skin. However, there is evidence from animal experiments that fluorocarbons can be absorbed through the skin while other experiments carried out using human skin samples have also shown that they can enter our bodies via the skin [21].

What Effect Do Fluorocarbons Have on Our Health and Wellbeing?

It is worth emphasizing that the amount of fluorocarbons in our bodies is very small. Based on the information that is available, most of us would have no more than a 100 millionths of a gram (ie about a tenth of a milligram) in our entire bodies [14]. Now the toxicity of PFOA, expressed as the LD50, has been estimated to be

250–500 mg per kilogram body weight of the rat. For PFOS, the LD50 is estimated to be about 200–300 mg per kilogram body weight. This means that, at least for rats, and assuming that a rat weighs 100 grams, you would need 25–50 mg of PFOA and 20–30 mg PFOS to kill most animals. A milligram is not very much. If a teaspoon of salt weighs about 6–7 gram then 25 milligrams is less than a two hundredth of a teaspoon. Of course, rats are not humans and it is likely that there are significant differences in toxicity between us and rats, so we do not know how much is required to kill humans. However, what it does illustrate is that what is in our bodies is almost certainly much less than that required to kill people outright.

We are rarely exposed to toxic doses of fluorocarbons so, for most of us, this is not an issue. However, based on what we now know, all of us are exposed to tiny – and apparently non-toxic amounts - every day of our lives. The actual daily dose we get almost certainly varies from person to person. Remember, too, that because the fluorocarbons we are exposed to are not just PFOA and PFOS but a mixture of many different substances, so not only does the dose we get vary but, so too does the composition of the fluorocarbon mixture vary. It is even more complicated than this because our age, sex, whether we drink or smoke, whether we have pre-existing medical conditons such as diabetes and heart disease, and our genetic make up, all probably influence our capacity to cope with our daily dose of fluorocarbons. These many variables make it very difficult to draw firm conclusions as the effects of fluorocarbon exposure on our health. With this in mind, let us look at what the latest scientific research tells us. To do this, we will look firstly at the effects of sub-lethal amounts of fluorocarbons on animals because studies on animals often, but not always, provide an indication of what is likely to happen in humans. Secondly, we will look at what the studies on humans tell us.

What Are the Effects of Sub-lethal Doses of Fluorocarbons on Animals?

Studies have been carried out to assess the effects of repeated small doses of PFOA and PFOS on rats and monkeys [45, 46]. PFOS seems to produce a greater effect at lower doses than PFOA [47]. Doses of 18 mg/kg body weight and above of PFOS, or 1.8 milligrams (or 1.8 thousands of a gram) for a 100 g rat per day for 90 days, can result in death [45]. This is less than a tenth of the toxic dose and illustrates the fact that repeated non toxic doses of a chemical can be harmful if taken long enough. As little as 10 mg per kg body weight given daily appeared sufficient to kill monkeys within the 90 day trial period. In one of the studies, the animals died at 5–7 weeks but before they died they were dehydrated, and had diarrhea and emesis (the medical term for vomiting). Post mortem analysis showed that there were changes in the liver and adrenals. In another study, even smaller amounts of PFOS (4.5 mg/kg body weight) given daily was sufficient to kill monkeys. As these sorts of abnormalities were also observed with a different species of monkeys, the cynomolgus, it is likely that all primates, including humans, would be affected in a similar way.

Of course, as mentioned above, people are exposed as well to many other fluorocarbons, not just PFOS and PFOA, and much less is known of their effects on animals, nor is there much information on the toxicity of the complex fluorocarbon mixtures we are exposed to.

What Are the Effects of Sub-lethal Doses on Us?

The conclusion to all of the studies on animals, including monkeys, is that chronic exposure to even very small doses of fluorocarbons, can be harmful. Still, while the doses that cause problems in animals are small, nevertheless they are much larger, perhaps by a factor of a thousand or more, than those we are exposed to so. Governments around the world have assured us that these amounts are too small to affect our health. If a millionth of a gram is a microgram, then, according to the UK Health Protection Agency (HPA) the tolerable daily intakes (TDI) ie the maximum amounts of PFOS and PFOA that can be ingested on a daily basis without an appreciable health risk, are 0.3 (three tenths) of a microgram of PFOS and 3 micrograms of PFOA per kilogram body weight respectively. The corresponding figures for the European Food Safety Authority are about half this. Taking the higher HPA figures, this means that, for a 70 kg male or female, daily intakes of PFOS and PFOA should not exceed 21 micrograms and 210 micrograms respectively.

Estimates for what we are actually exposed to are very difficult because, as discussed above, there are many sources of potential exposure – food, water, air, skin, cookware etc. We have an idea what is in the food that we eat but, even here, our exposure varies quite a lot according to our diet. However, it is generally believed that food is the major source of the fluorocarbons that are found in different parts of our bodies and there are some published figures for this. For example, if we take the figures reported for the Swedish diet, and assume that diets in other parts of the world are not that different, then the maximum daily dietary exposures for PFOS and PFOA are about one tenth and one twentieth of a microgram respectively. If the dietary figures are correct then the dietary intake of fluorocarbons is much lower than the TDI and therefore there is unlikely to be any impact on our health.

There are many caveats to all of this. Firstly, the TDI figures are really just an estimate and based on toxicology studies with monkeys and rats, not people. Allowance has been made for any unusual sensitivity we may have to these substances but even so, without actually carrying out tests on people, there is a degree of uncertainty. Secondly, these figures are based on studies carried out on PFOS and PFOA, not the complex mixture of fluorocarbons we are exposed to. It is possible, even likely, that the individual fluorocarbons may synergise with each other and produce quite unexpected effects at the levels we are exposed to. We will discuss synergy later in this book but suffice to say that the ability of individual chemical substances to produce much greater effects in combination with others is well known. Thirdly, while animals tested are mostly genetically similar, their diets carefully controlled, and are mostly are healthy without pre-existing conditions such as diabetes, obesity, or heart disease, the same is not true for humans. It is possible that some people, for whatever reason, are much more susceptible than others. Finally, most animal testing is for limited periods – there are exceptions to this of course – but our exposure to fluorocarbons is for life.

With these caveats in mind, and also because we cannot carry out the required studies on people, we have to resort to the next best thing – and that is to look at the results of epidemiological studies carried out on different groups of people and see if there is evidence at all that there is any indication that exposure to these apparently sub- TDI amounts can cause harm. We have discussed epidemiological studies elsewhere in this book but we should emphasise again that these types of studies are not unequivocal because there are a great number of variables that can influence to findings. In order to assess whether, for example, increased exposure to fluorocarbons can affect semen quality scientists would examine sperm from a few hundred or more men, determine what proportion of sperm are normal – there are lots of measurements that can be made – and also measure the type and levels of the various fluorocarbons in the blood of each individual. They would then look to see if any of the men had abnormal sperm and, if they did, whether they had higher amounts of fluorocarbons in their blood than the men who had normal sperm. If there appeared to be some sort of relationship between fluorocarbons in the blood and abnormal sperm the scientists would then have to carry out a statisticial analysis which is essentially a mathematical analysis of the data to determine the likelihood that the findings are not due simply to chance rather than to increased exposure to fluorocarbons. Even if it established that there is a strong statisitical relationship between blood fluorocarbons and sperm abnormality, it still does not prove that the fluorocarbons are the culprit, because there may be other factors that may not have been considered and therefore measured. All it does is indicate that there may be some relationship. On the other hand, and this is what often happens with this type of research, if many different groups of researchers find the same thing, then a relationship between fluorocarbons and sperm abnormality becomes much more likely. With all of this in mind, let us look at what has been found thus far.

Reproduction and Lactation

Semen Quality

Analysis of semen in young men has indicated that the amount of normal spermatozoa in semen is reduced significantly in those with the highest levels of fluorocarbons [22, 23, 42]. However, there is some uncertainty about this relationship because there have been other studies which have shown no effect at al [24].

Birth Weight

Birth weights of babies with higher levels of PFOS and PFOA in their blood or in their mother's blood have been reported to be lower. It has been suggested that prenatal exposure, through maternal blood, has an adverse effect on the growth of the developing baby [25, 26, 43]. Again, there is some inconsistency in the findings because some studies show little effect.

Breast Milk

As fluorocarbons are components of blood in most people, it is not entirely unexpected that these substances could find their way into human breast milk. In fact, analysis of breast milk from women in many different countries has confirmed the presence of many different types of fluorocarbons, the actual composition varying from country to country. As infant formula also contains fluorocarbons, most babies are exposed, some almost certainly more than others [27]. The unanswered question is – are they affecting our babies' health?

Heart Disease

Whether increased exposure to fluorocarbons such as PFOS and PFOA increase the risk of heart disease is open to debate. Some researchers have found that fluorocarbons such as PFOA can increase the risk while others have found that there is no effect [28, 31, 44]. Occupational exposure to refrigerant gases containing fluorocarbons such as Freon showed that exposure increased the risk of heart arrhymia (abnormal beating) [34]. There does appear to be an association between fluorocarbon exposure, as determined by measuring their levels in the blood, and increases in blood cholesterol, a risk factor in heart disease [48]. However, while the association seems to be consistent, the increases are very small and other factors such as body weight seem to be much more important [29, 30]. The effect on cholesterol is intriguing, particularly as the amounts of fluorocarbons involved are very tiny. It raises the very interesting question of how such small non toxic amounts of a chemical can produce changes in the levels of one of the key blood fats.

Kidney Disease

Our kidneys serve the important function of disposing of harmful waste products generated from the food we eat as well as environmental contaminants such as fluorocarbons. Abnormal kidney function is considered to be a risk factor for early death together with diseases of the heart and blood vessels. As fluorocarbons accumulate in significant amounts in the kidneys of laboratory animals and appear to produce pathological changes in the organ, there have been concerns that these substances may also affect the function of the kidney in humans [33, 50]. A study carried out on more than 4000 individuals found that those with higher blood levels of fluorocarbons were more likely to show signs of kidney abnormalities [32]. Of course this does not prove that there is an association between fluorocarbons and kidney disease in people for reasons that have been discussed earlier in this chapter. A more recent study has suggested that the changes in kidney function observed by some researchers are not caused by fluorocarbons but rather are are a consequence of an existing abnormality [48, 49]. More research is clearly needed.

Thyroid Disease

Our thyroid gland is a small organ situated in the front of the neck and produces two hormones, T3 and T4, which regulate the activities of many of the processes in our bodies. The activity of this organ is in turn regulated by TSH, a hormone produced by the brain. Changes in the levels of these hormones can result in disease. Studies with laboratory animals have shown that fluorocarbons can affect the function of the thyroid [35]. Fluorocarbons also may affect the function of the thyroid gland in humans because there are reports that there is an association between blood levels of fluorocarbons such as PFOA and PFOS, and TSH, T3 and T4 [50, 51]. These changes seem to trigger an iodine deficiency. Severe iodine deficiency states can lead to mental retardation and goiter (an enlargement of the gland which can affect swallowing). There is ongoing research into the possible effects of less severe iodine deficiency. There is a view that moderate iodine deficiency may increase the risk of preeclampsia (increased blood pressure) in preganant women and reduced intelligence in their babies. Interestingly, individuals with higher blood fluorocarbon levels have been reported to be more likely to have thyroid disease which seems to provide some support for the view that that fluorocarbons may affect thyroid gland function [35].

Gestational Diabetes

Diabetes is a complication of some preganancies (referred to as "gestational diabetes") with potential adverse effects on both the developing child and the mother. Little is known of the cause(s) of gestational diabetes although there are indications that certain unknown environmental factors may contribute to the development of the disease. Recent research suggests that fluorocarbons, and PFOA in particular, may be one of these enivironmental factors [52].

What Does All This Mean?

Fluocarbons are yet another group of chemicals produced by humankind that now have a ubiquitous presence throughout the environment being found in the soil, air, rivers, oceans, and wildlife literally everywhere on earth, even in areas far removed from human habitation. As with PCBs, PBDEs, organochlorine pesticides like DDT, plastics, heavy metals such as arsenic, lead and cadmium, and even the sweetener sucralose, their ubiquitous presence surely must lead to a questioning of the wisdom of releasing substances that are resistant to the normal degradative processes of our planet into the environment. Like the other resistant pollutants, fluorocarbons eventually their way into our food, water, and the air we breathe, and from there, into our blood and most of our organs. Indeed, there is now indisputable evidence that fluorocarbons are found in the blood and tissue of almost all humans, even those living in remote non-industrial parts of the planet eg Greenland.and the Faroe Islands.

What is also not in dispute is the fact that most people are exposed to only very small quantities of these pollutants, amounts that are much lower than those required to cause harm to a variety of different species of animals, including monkeys. Based on these studies, governments have set exposure limits and in almost all cases, even in those people exposed in an occupational setting where exposures are much greater than normal, there is no evidence that these limits are exceeded.

However, there are some grounds for concern because there are indications that these tiny amounts are not completely inert. In animals, what we do know is that sub lethal amounts of fluorocarbons, taken every day for an extended period of time, can cause disease and even death. As far as humans are concerned, there is a consistency in reports showing that greater exposure to fluorocarbons can lead to an increase in cholesterol in the blood, and this may explain why at least two groups of researchers have found that higher blood levels of fluorocarbons are associated with a slightly greater risk of heart disease. There are also indications- once again there is no absolute proof- that fluorocarbon exposure can affect the function of two other important organs, the kidney and the thyroid, and perhaps even reproductive processes such as sperm development and even birth weight.

As discussed repeatedly through this book, while it is relatively easy to determine the toxicity of a single chemical, many of the pollutants we are exposed to in our every day lives are complex mixtures and assessments of toxicity are rarely carried out on mixtures. The fluorocarbons that are taken up into our bodies are also mixtures of related substances, not just PFOA and PFOA, and we really know very little about what these mixtures do. And fluorocarbon mixtures are almost always found in the environment, and in our bodies, not alone but in combination with complex mixtures of other chemical pollutants such as PCBs, PBDEs, and organochlorine pesticides. The jury is still out on whether these mixtures of fluorocarbons, either by themselves, or in combination with mixtures of other pollutants may be having an impact on our health.

References

1. Zhao Z et al (2012) Distribution and long-range transport of polyfluoroalkyl substances in the Arctic, Atlantic Ocean and Antarctic Coast. Environ Pollut 170:71–77. http://www.ncbi.nlm.nih.gov/pubmed/22771353
2. Jahnke A et al (2007). http://www.ncbi.nlm.nih.gov/pubmed/17328178
3. Knobeloch L et al (2012) Perfluoroalkyl chemicals in vacuum cleaner dust from 39 Wisconsin homes. Chemosphere 88:779–783. http://www.ncbi.nlm.nih.gov/pubmed/22542201
4. Oono S et al (2008) Current levels of airborne polyfluorinated telomers in Japan. Chemosphere 73:932–937. http://www.ncbi.nlm.nih.gov/pubmed/18701130
5. de Solla SR et al (2012) Highly elevated levels of perfluorooctane sulfonate and other perfluorinated acids found biota and surface water downstream of an international airport, Hamilton, Ontario. Environ Int 39:19–26. http://www.ncbi.nlm.nih.gov/pubmed/22208739
6. Noorlander CW et al (2011) Levels of perfluorinated compounds in food and dietary intake of PFOA and PFOA in the Netherlands. J Agric Food Chem 59:7496–7505. http://www.ncbi.nlm.nih.gov/pubmed/21591675
7. Post GB et al (2012) Perfluorooctanoic acid (PFOA), an emerging drinking water contaminant: a critical review of recent literature. Environ Res 116:93–117. http://www.ncbi.nlm.nih.gov/pubmed/22560884
8. D'Hollander W et al (2010) Perfluorinated substances in human food and other sources of human exposure. Rev Environ Contam Toxicol 208:179–215. http://www.ncbi.nlm.nih.gov/pubmed/20811865
9. Vestergren R et al (2012) Dietary exposure to perfluoroalkyl acids for the Swedish population in 1999, 2005, and 2010. Environ Int 49:120–127. http://www.ncbi.nlm.nih.gov/pubmed/23018201
10. Eschauzier C et al (2013) Presence and sources of anthropogenic perfluoroalkyl acids in high consumption tap-water beverages. Chemosphere 90:36–42. http://www.ncbi.nlm.nih.gov/pubmed/22939265
11. Gellrich V et al (2013) Perfluoroalkyl and polyfluoroalkyl substances (PFASs) in mineral water and tap water. J Environ Sci Health A Tox Hazard Subst Environ Eng 48:129–135. http://www.ncbi.nlm.nih.gov/pubmed/23043333
12. Frisbee SJ et al (2009) The C8 health project: design, methods, and participants. Environ Health Perspect 117:1873–1882. http://www.ncbi.nlm.nih.gov/pubmed/20049206
13. Pirali B et al (2009) Perfluorooctane sulfonate and perfluorooctanoic acid in surgical thyroid specimens of patients with thyroid diseases. Thyroid 19:1407–1412. http://www.ncbi.nlm.nih.gov/pubmed?term=pfoa%20AND%20autopsy
14. Maestri et al (2006) Determination of perfluorooctanoic acid and perfluorooctanesulfonate in human tissues by liquid chromatographic/single quadrupole mass spectrometry. Rapid Commun Mass Spectrom 20:2728–2734. http://www.ncbi.nlm.nih.gov/pubmed/16915561
15. Schroter-Kermani C et al (2012) Retrospective monitoring of perfluorocarboxylates and perfluorosulfonates in human plasma archived by the German Environmental Specimen Bank. Int J Hyg Environ Health. http://www.ncbi.nlm.nih.gov/pubmed/22999890
16. Ericson I et al (2009) Levels of perfluorinated chemicals in municipal drinking water from Catalonia, Spain. Arch Environ Contam Toxicol 57:631–638. http://www.ncbi.nlm.nih.gov/pubmed/19685096
17. Shoeib M et al (2011) Indoor sources of poly- and perfluorinated compounds (PFCS) in Vancouver, Canada: implications for human exposure. Environ Sci Technol 45:7999–8005. http://www.ncbi.nlm.nih.gov/pubmed/21332198
18. Langer V et al (2010) Polyfluorinated compounds in residential and nonresidential indoor air. Environ Sci Technol 44:8075–8081. http://www.ncbi.nlm.nih.gov/pubmed/20925396
19. Beesoon S et al (2012) Exceptionally high serum concentrations of perfluorohexanesulfonate in a Canadian family are linked to home carpet treatment applications. Environ Sci Technol 46:12960–12967. http://www.ncbi.nlm.nih.gov/pubmed/23102093

20. Trudel D et al (2008) Estimating consumer exposure to PFOS and PFOA. Risk Anal 28:251–269. http://www.ncbi.nlm.nih.gov/pubmed/18419647
21. Franko J et al (2012) Dermal penetration potential of perfluorooctanoic acid (PFOA) in human and mouse skin. J Toxicol Environ Health A 75:50–62. http://www.ncbi.nlm.nih.gov/pubmed/22047163
22. Joensen-Nordstrom UN et al (2009) Do perfluoroalkyl compounds impair human semen quality. Environ Health Perpect 117:923–927. http://www.ncbi.nlm.nih.gov/pmc/articles/PMC2702407/
23. Toft G et al (2012) Exposure to perfluorinated compounds and human semen quality in Arctic and European populations. Hum Reprod 27:2532–2540
24. Raymer JH et al (2012) Concentrations of perfluorooctane sulfonate (PFOS) and perfluorooctanoate (PFOA) and their associations with human semen quality measurements. Reprod Toxicol 33:419–427. http://www.ncbi.nlm.nih.gov/pubmed/21736937
25. Lee YJ et al (2012) Concentrations of perfluoroalkyl compounds in maternal and umbilical cord sera and birth outcomes in Korea. Chemosphere. http://www.ncbi.nlm.nih.gov/pubmed/22990023
26. Chen MH et al (2012) Perfluorinated compounds in umbilical cord blood and adverse birth outcomes. PLoS One 7:e42474. http://www.ncbi.nlm.nih.gov/pubmed/22879996
27. Fujii Y et al (2012) Levels and profiles of long-chain perfluorinated carboxylic acids in human breast milk and infant formulas in East Asia. Chemosphere 86:315–321. http://www.ncbi.nlm.nih.gov/pubmed/22113060
28. Sakr CJ et al (2009) Ischaemic heart disease mortality study among workers with occupational exposure to ammonium perfluorooctanoate. Occup Environ Med 66:699–703. http://www.ncbi.nlm.nih.gov/pubmed/19553230
29. Frisbee SJ et al (2010) Perfluorooctanoic acid, perfluorooctanesulfonate, and serum lipids in children and adolescents: results from the C8 health project. Arch Pediatr Adolesc Med 164:860–869. http://www.ncbi.nlm.nih.gov/pubmed/20819969
30. Kerger BD et al (2011) Tenuous dose response for common disease states: a case study of cholesterol and pefluorooctanoate/sulfon ate (PFOA/PFOS) in the C8 health project. Drug Chem Toxicol 34:396–404. http://www.ncbi.nlm.nih.gov/pubmed/21770727
31. Shankar A et al (2012) Perfluorooctanoic acid and cardiovascular disease in US adults. Arch Intern Med. http://www.ncbi.nlm.nih.gov/pubmed/22945282
32. Shankar A et al (2011) Perfluoroalkyl chemicals and chronic kidney disease in US adults. Am J Epidemiol 174:893–900. http://www.ncbi.nlm.nih.gov/pubmed/21873601
33. Cui L et al (2009) Studies on the toxicological effects of PFOA and PFOS on rats using histological observation and chemical analysis. Arch Environ Contam Toxicol 56:338–349. http://www.ncbi.nlm.nih.gov/pubmed/18661093
34. Sabik LM et al (2009) Cardiotoxicity of Freon among refrigeration services workers: comparative study. Environ Health 8:31. http://www.ncbi.nlm.nih.gov/pubmed/19594908
35. Melzer D et al (2010) Association between serum perfluoroctanoic acid (PFOA) and thyroid disease in the US National Health and nutrition examination survey. Environ Health Perspect 118:686–692. http://www.ncbi.nlm.nih.gov/pmc/articles/PMC2866686/
36. Yu N et al (2015) Distribution of perfluorooctane sulfonate isomers and predicted risk of thyroid hormonal perturbation in drinking water. Water Res 76:171–180
37. Gebbink WA et al (2015) Perfluoroalkyl acids and their precursors in Swedish food: the relative importance of direct and indirect dietary exposure. Environ Pollut 198:108–115
38. Ericson I et al (2018) Levels of perfluorochemicals in water samples from Catalonia, Spain: is drinking water a significant contribution to human exposure? Environ Sci Pollut Res Int 15:614–619
39. Perez F et al (2013) Accumulation of perfluoroalkyl substances in human tissues. Environ Int 59:354–362
40. Rice PA (2015) C-6 Perfluorinated compounds. The new greaseproofing agents in food packaging. Curr Environ Health Rep 2:33–40

41. Sinclair E et al (2007) Quantitation of gas-phase perfluorolkyl surfactants and fluorotelomer alcohols from non-stick cookware and microwave popcorn bags. Environ Sci Technol 41:1180–1185
42. Louis GM et al (2015) Perfluorochemicals and human semen quality: the LIFE study. Environ Health Perspect 123:57–63
43. Bach CC et al (2015) Perfluoroalkyl and polyfluoroalkyl substances and human fetal growth: a systematic review. Crit Rev Toxicol 45:53–67
44. Mattsson K et al (2015) Levels of perfluoroalkyl substances and risk of coronary disease: findings from a population-based longitudinal study. Environ Res 142:14152
45. COT statement on the tolerable daily intake for ammonium perfluorooctanate. Committee on Toxicity of Chemicals in Food. October 2006
46. Butenhoff et al (2002) Toxicity of ammonium perfluorooctanate in male cynomolgus monkeys after oral dosing for 6 months. Toxicol Sci 69:244–257
47. Public Health England report. PFOS and PFOA: Overall toxicological overview
48. Zeng XW et al (2015) Association of polyfluoroalkyl chemical exposure with serum lipids in children. Sci Total Environ 512–513:364–370
49. Watkins DJ et al (2013) Exposure to perfluoroalkyl acids and markers of kidney function among children and adolescents living near a chemical plant. Environ Health Perspect 121:625–630
50. Kataria A et al (2015) Association between perfluoroalkyl acids and kidney function in a cross sectional study of adolescents. Environ Health 14:89
51. Jain RB et al (2013) Association between thyroid profile and perfluoroalkyl acids: data from NHNAES 2007-2008. Environ Res 125:51–59
52. Zhang C et al (2015) A prospective study of prepregnancy serum concentrations of perfluorochemicals and risk of gestational diabetes. Fertil Steril 103:184–189
53. Becanova J et al (2016) Screeing for perfluoroalkyl acids in consumer products, building materials, and wastes. Chemosphere 164:322–329
54. Wang Z et al (2015) Hazard assessment of fluorinated alternatives to long-chain perfluoroalkyl acids (PFAAs) and their precursors:status quo, ongoing challenges and possible solutions. Environ Int 75:172–179
55. US EPA Emerging Contaminant Fact Sheet March 2014. Emerging contaminants -Perfluorooctane sulfonate (PFOS) and Perfluorooctanoic Acid

Chapter 11
Noise, Chemicals, and Hearing

The sounds of waves breaking on a deserted beach, bird song, the croaking of frogs, rain, and gentle breezes rustling the leaves of trees, are all soothing to the soul. Not so soothing are other natural sounds such as lightning and thunder, volcanic eruptions, cyclonic winds and large hailstones falling to earth. While there are some sounds produced by humankind are soothing, for example the exquisite melodies of Beethoven, Tchaikovsky and Mozart, many of the sounds, particularly those in big cities, are jarring, stressful and, in some instances, almost painful. The roar of giant trucks, motorcycles, jackhammers, and train horns are hardly soothing and, unlike most natural sounds are frequent and continuous, particularly in big cities.

What Is Sound?

Sound is produced from the vibration of the molecules in air (mainly oxygen, nitrogen and carbon dioxide). The vibrating air molecules generate sound waves of varying frequency, the higher the frequency the higher the "pitch". The sound waves eventually reach the ear, a highly complex structure, where their energy is converted into nerve impulses which in turn are carried via the auditory nerve to specific sites within the brain. The latter then interpret these signals as "sounds". The ear itself is highly complex and divided into outer, middle, and inner ear and associated structures such as the tympanic membranes, ossicles, and cochlea. Not only is the ear essential for hearing, through special canals containing fluid and hair-like sensors it is involved in the maintenance of balance. As in every other part of the body, the chemical composition of the ear and its associated components is even more complex containing scores of special proteins, fats, carbohydrates and a variety of other substances each with its own unique function. It is easy to see how interference with this exquisitely organized structure in any number of ways can result in hearing loss and the often associated tinnitus (ringing in the ears) and/or balance problems.

© The Author(s), under exclusive license to Springer Nature Switzerland AG 2021   157
A. Poulos, *The Secret Life of Chemicals*,
https://doi.org/10.1007/978-3-030-80338-4_11

How Is Sound Measured?

The intensity of a sound is measured is measured in units termed "decibels", the greater the number of decibels the more intense the sound. The hearing threshold for the human ear is 0 decibels, while whispering, normal conversations and loud conversations produce around 30, 60, and 70 decibels respectively. Sounds produced by jet engines (at 30 metres distance), chain saws, train horns, lawn mowers, and heavy kerbside traffic are around 140, 110, 100, 90, and 80 decibels respectively [1]. It is important to note that as far as the effect of sound on the human ear is concerned there is an additional factor ie the duration of the sound. If the sound produced is unpleasant, harsh, or intermittent it is normally classified as noise. There is also a subjective element to the definition of noise because what is considered noise by one person may thought to be a pleasant "sound" and not noise by someone else. The different genres of music are good examples of this.

Sources of Noise

It is fairly obvious that, particularly in cities, noise is everywhere. We are exposed to noise even through every day simple activities, such as catching a tram, train or bus, or even walking along a busy street, or being close to a building site. Then there are social activities such as the cinema, and attending classical, rock or jazz concerts, and crowded restaurants and coffee shops. Increasingly, noise exposure also occurs through devices such as i-Pads or telephones, but in this case the exposure occurs via ear phones. And, of course, many people are exposed to varying amounts of noise in their work environments.

Noise Pollution

According to the generally accepted definition, the word pollution refers to contamination with poisonous or harmful substances. In the case of noise pollution, there is a "contamination" of the environment with noise rather than poisonous or harmful substances. As mentioned above, what constitutes "noise" is subjective so the definition of what constitutes noise pollution is not straight forward. However, for the purposes of determining the impact of noise pollution on our health and wellbeing, there are two principal components – loudness which refers to the numbers of decibels of a particular sound, and the length of time exposed to the sound. Of course there is another component, which is the frequency of the sound which determines whether the noise is high or low pitched.

The Health Effects of Noise

Hearing Loss

As in other part of the body, the ear and its various components are subject to gradual decay and the net result of this is hearing loss. For example, it is estimated that around a third of people in the USA over the age of 65 have significant hearing loss [2]. It is believed that there are many factors, for example obesity, diet, smoking, and genetics that contribute to hearing loss in older people [3]. Noise is another factor because it has been reported that up to a quarter of adults in industrialised countries such as the USA may suffer from work related hearing loss as compared to those living in countries with a relatively noise free environment where the prevalence of hearing loss is lower [2, 4]. It is likely that this is not unique to the USA but is a feature of most industrialised countries [1–3].

For many people an important factor in their hearing loss is their work environment. It has been estimated that noise induced hearing loss is a common occupational disease in major industrialised countries and that in countries like Australia and the USA perhaps as many as almost one million and thirty million respectively may be exposed to levels of noise that could increase the risk of hearing loss [5, 6]. It is believed that one of the major causes of noise-induced hearing loss is the effect of excessive noise on the delicate hair cells within the cochlear, a key part of the ear which plays a role in the conversion of sound to the nerve impulses which travel to the brain.

The integration of music into personal listening devices, particularly mobile phones, has greatly increased the exposure of people to varying intensities of sound. Preferred volume levels can vary greatly amongst listeners. Perhaps not surprisingly there is some evidence that there are reductions in hearing in those individuals who choose to listen to music at high volume levels [7].

Cardiovascular Disease

Cardiovascular disease, ie disease affecting the heart and blood vessels, is the most common cause of death in the world [8]. It has been estimated that in 2016 around 17.9 million people, or almost a third of all global deaths, died from cardiovascular disease. There are numerous risk factors including diet, salt intake, alcohol consumption, smoking, and physical inactivity. Over the last few years there have been reports linking occupational exposure to noise with an increased risk of cardiovascular disease [9, 10]. There are also indications of a slightly increased risk of heart disease with noise pollution even in a non occupational setting from aviation, rail, and road traffic [11–13]. It has been suggested that environmental noise causes a release of stress hormones and inflammatory chemicals which in turn may produce effects on the heart and blood vessels and this may explain the possible link with heart disease [14].

Sleep

Sleep is an important physiological function and necessary for good health and wellbeing. The view of the American Academy of Sleep Medicine and Sleep Research Society is that insufficient sleep may have potentially harmful effects on cardiovascular, metabolic, and mental health, as well as possible effects on the immune system, and the development of cancer [15]. It follows that anything that interferes with sleep may have detrimental effects on our health and wellbeing. There are many reasons for sleep disturbance and one of these is environmental noise, particularly from road and rail traffic and from aircraft [16].

Weight Gain

Weight gain is now a major health issue in all developed countries because of its possible links to cardioavascular disease and diabetes. There is evidence that exposure to certain types of noise, for example road traffic noise, may increase the risk of obesity [17, 18]. It has been suggested that changes in body weight from noise exposure could be the result of hormonal and physiological responses to stress and sleep disturbance [19].

Acoustic Neuroma

Acoustic neuroma is a condition caused by the growth of a non-malignant tumor on one of the cranial nerves, specifically the one that links the inner ear to the brain. There are indications that continuous and excessive noise may increase the risk of this disease [20].

Tinnitus

There is another effect of noise pollution on the normal functioning of the ear, and that is tinnitus. Tinnitus is a ringing or buzzing in the ears and often associated with hearing loss. The condition can affect either one ear or both. While age and genetic factors contribute to the development of tinnitus, there clear evidence that occupational exposure to noise, and perhaps to a lesser degree, even leisure noise, such as music, can contribute [21, 22].

Hearing Loss and Chemicals

Most people would be surprised to learn that there is another cause of hearing loss and that is exposure to what are termed "ototoxic" chemicals ie chemicals that act as poisons to the ear. Effects may be reversible once the exposure has ceased or irreversible. The chemicals include some of the therapeutic drugs, in particular platinum containing drugs cisplatin, used as chemotherapeutic agents to treat cancer, antibacterial agents for treating infections eg aminoglycosides such as streptomycin, gentamycin, neomycin, and others found in the workplace eg pesticides, heavy metals such as lead, cadmium and mercury, solvents, and some chemicals used in the manufacture of plastics eg styrene [23, 24]. Certain drugs may also induce tinnitus [25]. There are also suggestions that exposure to certain agricultural chemicals, and even some environmental pollutants such as polychlorinated biphenyls, the metals cadmium and lead, and organochlorines can also cause hearing loss [26–30]. It is thought that these chemicals produce their effects by acting on one for more of the key components of ears, in particular the cochlear, the auditory nerve, or the vestibular system.

While noise and chemical exposure can each independently cause hearing loss, there is evidence that simultaneous exposure to noise and chemicals such as solvents eg xylene, toluene, or styrene, or metals, such as cadmium or lead, as may occur in an industrial setting may damage hearing more than exposure to noise or chemicals alone [23, 24].

As genetic factors are important in the development of many diseases, it is not surprising that there has been considerable research into determining whether our genes play a role in hearing loss resulting from exposure to noise and chemicals [31]. As has been observed for many disease states, that there is also an increasing amount of research into the role of inflammatory processes in the development of hearing loss [32, 33]. It has been speculated that cisplatin and some of the ototoxic antibiotics are taken up into the cochlear and trigger inflammatory changes which ultimately lead to hearing loss. Another mechanism that is being investigated is the possible role of free radicals, highly reactive substances formed during normal metabolic processes, and regulated naturally in the body by antioxidants such as vitamins C and E [33].

What Does This Mean?

There is no question that those of us who live in cities are exposed to constant noise – from motor vehicles, trains, and aircraft as well as from the myriad activities that are a normal part of city life. Many of us are also exposed to noise through leisure activities such as concerts, movies, television, and mobile phones, although the noise that is of greatest concern is that which occurs in an occupational setting. While brief exposure to loud noise is not normally harmful, unless the volume of noise is excessive such as, for example, that which occurs in close proximity to an explosion, If the noise is both loud and continuous for extended periods such as days or years, hearing loss can result. For many people this may occur in their working environment. If, at the same time, there is exposure to certain chemicals the risk of hearing loss is increased. Most of the so-called ototoxic chemicals identified thus far that have been shown to affect hearing are pharmaceuticals like, for example, cisplatins and some antibiotics. However, it has become increasingly clear that a variety of other chemicals, for example certain solvents and metals, polychlorinated biphenyls and organochlorines, can also affect our hearing. Little is really known of the effects on our hearing of many of the other chemical pollutants present in the air, water, and food, particularly in cities where exposure can take place in noisy environments over long periods of time. There is preliminary evidence that at least some of these pollutants, for example, the metals lead and cadmium, certain pesticides, and others such as PCBs, may damage the delicate structures within the ear. Confirmation of any deleterious effects of the scores of environmental chemicals on our hearing is exceedingly difficult because they are likely to occur gradually over long periods of time. However, it is tempting to speculate that, at least in susceptible individuals, the combined effects of pollution and noise in city environments, can cause hearing loss.

References

1. Safework Australia. https://www.safeworkaustralia.gov.au/noise. Noise
2. Liberman MC (2017) Noise-induced and age-related hearing loss: new perspectives and potential strategies. F1000Res 6:927
3. Lie A et al (2016) Occupational noise exposure and hearing: a systematic review. Int Arch Occup Environ Health 89:351–372
4. Counter SA, Klareskov B (1990) Hypoacusis among the Polar Eskimos of Northwest Greenland. Scand Audiol 19:149–160
5. Stucken E (2014) Noise-induced hearing loss: an occupational medicine perspective. Curr Opin Otolaryngol Head Neck Surg 22:388–393
6. Safework Australia. Work related hearing loss in Australia. https://www.safeworkaustralia.gov.au/doc/work-related-noise-induced-hearing-loss-australia
7. Husssain T et al (2019) Early indication of noise-induced hearing loss in young adult users of personal listening devices. Ann Otol Rhinol Laryngol 127:703–709
8. World Health Organisation (2017, May) Cardiovascular diseases (CVDs). https://www.who.int/news-room/fact-sheets/detail/cardiovascular-diseases-(cvds)
9. Eriksson HP et al (2018) Longitudinal study of occupational noise exposure and joint effects with job strain and risk for heart disease and stroke in Swedish men. BMJ Open 8:e019160
10. Yang Y et al (2018) Relationship between occupational noise exposure and the risk factors for cardiovascular disease in China: a meta-analysis. Medicine (Baltimore) 97:e11720

11. Seidler A et al (2016) Aircraft, road, and traffic noise as risk factors for heart failure and hypertensive disease. A case control study based on secondary data. Int J Hyg Environ Health 219:749–758

12. Kempen EV et al (2018) WHO environmental noise guidelines for the European region: a systematic review of environmental noise and cardiovascular and metabolic effects: a summary. Int J Environ Res Public Health 15(2):pii;E379

13. Peters IL et al (2018) Aviation noise and cardiovascular health in the United States: a review of the evidence and recommendations for future directions. Curr Epidemiol Rep 5:140–152

14. Munzel T et al (2018) Environmental noise and the cardiovascular system. J Am Coll Cardiol 71:688–697

15. Grandner MA (2017) Sleep, health and society. Sleep Med Clin 12:1–22

16. Perron S et al (2016) Sleep disturbance from road traffic, railways, airplanes and from total environmental noise levels. Int J Environ Res Public Health 13:pii E809

17. Foraster M et al (2018) Long term exposure to transportation noise and its association with adiposity markers and development of obesity. Environ Int 121:879–889

18. Oftedal B et al (2015) Road traffic noise and markers of obesity – a population based study. Environ Res 138:144–153

19. Muscogiuri G et al (2019) Obesity and sleep disturbance: the chicken or the egg? Review Crit Rev Food Sci Nutr 59:2158–2165

20. Cao Z et al (2019) Noise exposure as a risk factor for acoustic neuroma: a systematic review and meta-analysis. Meta-Analysis Int J Audiol 58:525–532

21. Ralli M et al (2017) Work-related noise exposure in a cohort of patients with chronic tinnitus. Analysis of demographic and audiological characteristics. Int J Environ Res Public Health 14: pii:E1035

22. Moore DR et al (2017) Lifetime leisure music exposure associated with frequency of tinnitus. Hear Res 347:18–27

23. Hormozi M et al (2017) The risk of hearing loss associated with occupational exposure to organic solvents mixture with and without concurrent noise exposure: a systematic review and meta-analysis. Int J Occup Med Environ Health 30:521–525

24. Choi YH, Kim K (2014) Noise-induced hearing loss in Korean workers: co-exposure to organic solvents and heavy metals in nationwide industries. PLoS One 9:e97538

25. Lanvers-Kaminsky C et al (2017) Drug-induced ototoxicity: mechanisms, pharmacogenetics, and protective strategies. Review Clin Pharmacol Ther 101:491–500

26. Xu H et al (2020) Association between pyrethroid pesticide exposure and hearing loss in adolescents. Environ Res 187:109640

27. Min JY et al (2014) Serum polychlorinated biphenyls concentration and hearing impairment in adults. Chemosphere 102:6–11

28. Rosati R, Jamesdaniel S (2020) Environmental exposures and hearing loss. Int J Environ Res Public Health 17:4879

29. Choi YH et al (2012) Environmental cadmium and lead exposures and hearing loss in U.S adults: the National Health and nutrition Examinatiion survey, 1999-2004. Environ Health Perspect 120:1544–1550

30. Zhang J et al (2021) Environmental exposure to organochlorine pesticides and its association with the risk of hearing loss in the Chinese adult population. A case-control study. Sci Total Environ 767:145153

31. Miao L et al (2019) An overview of research trends and genetic polymorphisms for noise-induced hearing loss from 2009-2018. Environ Sci Pollut Res Int 26:34754–34774

32. Frye M et al (2019) Inflammation associated with noise-induced hearing loss. J Accoust Soc Am 146:4020–4032

33. Yu D et al (2020) Current strategies to combat cisplatin-induced ototoxicity. Front Pharmacol 11:999

Chapter 12
Radiation Pollutants

Every day of our lives we are bombarded with radiation, in one form or another. There are two main forms of radiation, the so-called non-ionising and ionizing radition. Radiation is simply a form of energy. The term "ionizing" refers to the ability of some forms of radiation to strip one or more of the electrons surrounding the nucleus of atoms resulting in the formation of atoms with a positive electric charge. These are termed "ions", hence word "ionizing". The best known example of this form of radiation is radioactivity which is produced by chemical substances such as uranium. We will discuss radioactivity in greater detail later in this chapter. The energy from radiation is released mostly as either electrical particles e.g., electrons (negative electrical charge) and protons (positive electrical charge), or electromagnetic waves. These particles or waves are able to travel through space and, if they collide with matter, such as for example, human tissues, they can strip electrons from atoms causing damage. Of greatest concern is their ability to cause changes in the genetic material present in most of the tissues in our bodies.

The cosmic rays that shower our planet every day come from outside our solar system and are mostly atoms stripped of electrons and are a form of ionizing radiation. These particles, referred to as "primary" cosmic rays interact with gases in the Earth's atmosphere generating "secondary" cosmic rays that are mostly sub atomic partcles such as mesons.

Non ionizing radiation is comprised of electromagnetic waves and includes visible and infrared light, longer wave length ultraviolet light (UV-A and UV-B), microwaves, and radiowaves. Some types of ultraviolet light (far UV and extreme UV) are generally included in the ionizing radiation category because their high energy content can cause ionization of whatever medium it passes through. While non ionizing radiation has less energy and is therefore mostly not able to strip electrons from atoms, it may induce temperature changes as it passes through material such as human tissues and this can cause damage. It is also believed that there may be other, more subtle, or non thermal effects. While the thermal effects of non ionizing radiation are reasonably well understood, little is really known about what these non thermal effects are and whether they are really harmful.

© The Author(s), under exclusive license to Springer Nature Switzerland AG 2021
A. Poulos, *The Secret Life of Chemicals*,
https://doi.org/10.1007/978-3-030-80338-4_12

What Are the Natural Sources of Ionizing Radiation?

There are many natural sources of ionizing radiation. Rock and soil contain radio-active elements (discussed later), and many radioactive elements are found in trace amounts in the oceans and in other surface waters [1, 2, 33]. They are also released during volcanic eruptions and from geothermal activity [3, 34]. Cosmic rays have already been mentioned with exposures at high altitudes, for example in aircraft, estimated to be many times greater than at ground level [4]. Most of the ultraviolet light reaching the surface of the Earth is the non ionizing UV-A and B.

The different forms of light including visible light, UV- A and B, and infrared light are all natural sources of non ionizing radiation. Radio waves released by interstellar objects and reaching the Earth are yet another form, and non ionizing radiation is also generated by lighting strikes.

What Is Radioactivity?

All matter is made up of elements and there are at least 98 naturally occurring elements found in nature. Hydrogen, oxygen, and nitrogen are a few examples of elements. The elements themselves in turn are made up of atoms, which contain varying numbers of protons (with a positive electrical charge), neutrons (no electric charge) and electrons (negative electrical charge). The simplest atom is hydrogen which consists of one proten, and one electron, while oxygen has eight protons, eight neutrons and eight electrons. Most of the naturally occurring atoms that make up matter are stable, that is they do not degrade. On the other hand, the atoms of some elements, and uranium is a good example, degrade or, to use the scientific term, decay. Uranium is one of many radioactive elements. It exists in different forms, and these forms are termed "isotopes". Uranium 235 and uranium 238 are two isotopes of uranium which decay releasing alpha particles. The latter carry an electric charge, have high energy content, and can therefore cause ionisation, This release of alpha particles continues almost indefinitely. In the case of uranium 235 and uranium 238 it takes thousands of years until, finally, the uranium atoms have been converted into forms of lead (lead 207 and lead 206 respectively). The measure of how long it takes for a radioactive atom to decay is the half life, which is the time taken for half of an amount of the radioactive element to decay. Like uranium, all radioactive elements decay into some other element.

The half lives, and hence how quickly the various radioactive elements decay varies greatly – from the millions of years for some isotopes of uranium, more than 5000 years for carbon 14 (used in medical research), about 28–30 years for cesium 137 and strontium 90, 87 days for sulphur 35, and 8 days for iodine 131. The half lives of some radioisotopes are extremely short, decaying in fractions of a millisecond (thousandths of a second).

We have already mentioned the release of alpha particles, or electrically positive charged particles, but some radioactive elements release electrons (negatively charged particles), positrons (which have the same particle size as electrons but have a positive charge) or gamma radiation. Gamma radiation is a form of light with a very high energy content.

When the various particles are released from radioactive atoms they travel through air or whatever medium they are contained in. Alpha particles travel only a few centimetres in air, and can be stopped by a sheet of paper, and do not normally penetrate the skin. Beta particles travel metres through the air, can be stopped by a thin layer of plastic or metal, and penetrate a few millimetres through the skin. Gamma rays also move a considerable distance from their source, and can only be stopped by a layer of dense material such as lead or concrete or metres of water. They can penetrate the skin and even pass through the body.

What Are the Principal Radioactive Elements Used by Humankind?

There is little doubt that the word "radioactive" generates much fear in human beings. Images of atomic bomb blasts, the Cold War and the threats of nuclear war, and disasters in nuclear power plants such as Chernobyl and Fukushima, all contribute to this fear. However, there is another aspect of radiaoactivity that is not fully understood, much less appreciated. While some of the radioactive elements can be used to create the chaos and destruction of a nuclear war, others are being used for the overall good of humankind. There are over 200 radiosotopes that have found uses in medicine, industry and research, and many of these do not occur naturally. Some of the radiosotopes used include –

- Iodine 131 (medicine)
- Strontium 90 (energy source)
- Cesium 137 (industry, medicine)
- Cobalt 60 (food irradiation, radiotherapy, industry)
- Technetium 99 (medicine)
- Samarium 153 (medicine)
- Polonium 238, 239, 240 (space probes, antistatics)
- Americium 241 (industry, medicine)
- Carbon 14 (research)
- Sulphur 35 (research)
- Tritium (nuclear weapons, research, luminescent devices)

What Are the Sources of Radioactivity Released into the Environment by Humankind?

Because small amounts of radioactivity are found in many minerals and rocks, it follows that the waste generated by mining will release radioactivity into the environment. Uranium mining produces the most radioactive waste but other mining such as aluminium, copper, silver, rare earths, gold, and rock phosphate (mined for fertilisers production) also generates radioactive waste [5]. Power generation, for example coal-fired plants, and oil refining, can also release radioactivity which is present in oil and gas. And, of course, all nuclear power plants generate radioactive waste, as do nuclear submarines [6]. Of course, there was the atomic bombing of Hiroshima and Nagasaki in 1945 and then there have been a number of accidents that have occurred over the last 50 years. One of the first was at the Kyshtym nuclear processing site in Ozyorsk, in what was a part of the USSR, in 1957. In 1979, radioactive material was released from the Three Mile Island power plant in Pennsylvania, USA, the Chernobyl plant in the Ukraine, then part of the USSR in 1986, and more recently the nuclear power plant in Fukushima, Japan in 2011. In addition to this, until relatively recently, various countries have carried out atomic testing, both underground and above ground, which also led to the generation and release of radioactive pollutants. A list of some of the activities that may result in a release of radioactivity into the environment is shown below

- Nuclear fuel production and reprocessing
- Nuclear power production
- Research
- Defence
- Production of radioactive chemicals for industry and medicine
- Medical and industrial usage
- Landfills and waste disposal
- Mining
- Coal and oil fueled power generation
- Nuclear accidents (e.g., Fukushima, Chernobyl)
- Atomic testing

Does Our Food and Water Contain Radioactive Substances?

Most people would be surprised to learn that much of the food we eat contains small amounts of radioactivity [7–9]. This is not surprising because soil, fertilisers, and even the water used for irrigation and drinking contain radioactive elements [10–12]. Some of this radioactivity, and it can vary greatly, comes from natural sources. Many of the elements which make up all animate and inanimate objects on the Earth exist is different isotopic forms, and some of these forms are radioactive. So, for example, there are 100 grams or more of potassium in our bodies and this comes from the food that we eat. Around 0.01% of this potassium exists as the radioactive form, potassium 40 (the normal isotope is potassium 39). So, any food that is rich in potassium – bananas are an oft quoted example- will contain small amounts of radioactivity due to potassium 40. Carbon 12 is the natural form of carbon and is present in all living plants and animals. Carbon 14 is a radioactive form and, while the natural abundance of this form of carbon is very low (about 1 in a trillion), nevertheless because of the fact that carbon it is present in all living things in large amounts (the human body contains over 20% carbon) it constitutes a significant component of the natural radioactivity of food. There are also areas around the world with high background radiation due to the presence of radioactive elements in the soil and rocks, and food grown in the area can contain much higher than normal levels of a variety of radioactive elements including radium 226, thorium 232, cesium 137 and uranium 238.

In addition to the naturally occurring elements, food may contain a variety of other radioactive substances that are products of human activities. A comprehensive study of the different radisotopes used by industry and medicine, their release into the environment, and their presence in different foods was carried out by the UK Environment Authority and a report issued in 2011. Some of these have been detected in a variety of different foods and water. According to the above mentioned report, people who live in different parts of the UK consume less than three hundredths of a millisievert per year in their drinking water and this does not appear to vary greatly even in areas that are close to nuclear reactors. The absolute amounts of radioactivity consumed in food and water in most regions of the UK were estimated to be less than two tenths of a millisievert, which is much less than the maximum level recommended by the various agencies of government (1 millisievert). Most of the radioactivity was due to the potassium 40 and carbon 14. While the natural radioactivity levels vary considerably around the world, it seems unlikely that levels are much greater than this anywhere in the world.

What Are the Sources of Non Ionizing Radiation?

The last decade or so has seen a proliferation of non ionizing radiation sources [32]. This type of radiation is generated from electric power lines, microwave ovens, mobile telephone towers, cordless telephones, TV and radio stations, medical and chemical devices, heating and tanning beds, and wind turbines to name just a few. Indeed, just about any equipment utilizing electricity or magnetic fields generates some sort of non ionizing radiation. Another source of non ionizing radiation is the ubiquitous laser. Lasers, which are essentially a form of high energy light, are used in electronic equipment such as computers, in industry and medicine.

What Are the Effects of Radioactivity?

As beta particles or gamma rays produced by radioactive substances penetrate tissue, they have the capacity to react with different chemical components of the cells that make up the tissue. The high energy can damage cells by reacting with, and damaging, some of the proteins, fats, and even the deoxyribonucleic acid (DNA) which make up our genes. Our cells have an enormous capacity for healing and regeneration, but if the damage is too extensive, the cells may die and the function of the organ impaired. High dose exposure can lead to damage to blood cell components, fatigue, nausea, vomiting and, eventually, death. Depending on the dose of radioactivity, death can occur fairly quickly or may take weeks or months.

A Few Words About Radioactive Dose

Mention should be made about dosage in any discussion of the health effects of radioactivity. There are two factors that are important in determining whether a dose of radioactivity is likely to harm- how much energy there is in the dose and the type of energy. Obviously, the amount of energy that is absorbed by the body or its organs, is important. The unit of measurement that describes how much radiation is absorbed is the gray – abbreviated to Gy. However, there are very significant differences in the degree of harm caused by the absorption of the different types of radiation released by a radioactive element. For example, alpha particles are more harmful than beta particles even if the amount of energy absorbed by the body is the same. So, to determine the potential for harm, the actual amount of energy absorbed from a particular exposure i.e., the number of Gys is multiplied by what is termed a "weighing factor". This measurement, which is called the "equivalent dose", is what is used by scientists as a means of assessing how potentially harmful a dose of radiation is likely to be. The weighing factor used for alpha particles is 20 while the corresponding factors for beta particles, x-rays and gamma ray is only one so a dose of alpha particles is potentially 20 times more potentially harmful than the corresponding dose of beta particles. The unit of measurement for an effective dose of radioactivity used by the USA is the "rem", while the "sievert" (or Sv) is used by most other countries. To convert rem to sieverts you need to divide the number of rem by 100.

Radioactivity Dosage and Health – Large Amounts

It is believed that full body exposure to around 8 Sv (800 rem) will kill all humans, while 3–6 Sv will kill half of those exposed. On the other hand, exposure of a single organ to radioactivity depends on the organ that is exposed with some organs, most notably those parts of the body, such as the blood cells and the sex cells which turn over the most, are much more sensitive than, for example, the skin where turnover is normally very slow. Exposure of single organs to relatively large doses of radioactivity is used therapeutically to kill cancers of certain organs, especially cancers of the thyroid and prostate glands. However, special care is required to ensure that surrounding normal tissue is not exposed, otherwise it too are killed by the radioactivity.

We can get an idea of what happens when people are exposed to high doses of radioactivity from a number of nuclear accidents that have occurred over the years.

The nuclear accident that occurred in Chernobyl in 1986 which resulted in the exposure of emergency workers to very large doses of radioactivity is a good example. Fatality rates for emergency workers that received large doses (above 6 sieverts) approached 95% and symptoms included nausea, vomiting, headaches, dizziness, disorientation, and fever [13]. One very important effect with serious consequences was the damage to the bone marrow, the source of the blood white and red cells. Similar effects were observed in those who survived the atomic bombs in Nagasaki and Hiroshima.

Radioactivity Dosage and Health – Small Amounts

While there is no doubt that high doses of radioactivity are harmful, and there is plenty of evidence to support this, the effects of lower dose exposures are much more complex [31]. Smaller, non lethal doses, of radioactivity can increase the risk of harm, but there are all sorts of factors that can either increase or decrease the risk. Some of these include –

- The dose of radiation
- The type of radiation (i.e., alpha, beta, or gamma)
- Duration of exposure (i.e., short period or chronic)
- Type of exposure (i.e., external whole body or internal)
- The organ that is exposed
- Age of person exposed
- Exposed person's health
- Source of exposure (i.e., air, water, food, skin)
- Genetic differences

With all of these variables, it is really not surprising that it is not that easy to draw any firm conclusions. There have been lots of studies that have examined this question, and many of these have been based on accidental exposure from nuclear accidents, the Japanese atomic bomb survivors, workplace exposure e.g., mining, and medical procedures e.g., X-rays, radiotherapy.

Dose and type of radiation is clearly an important factor. However, there are as many as five separate models that have been developed to try to explain the effects of small doses [14]. The two main ones are the so-called "linear no threshold" (LNT) and the "radiation hormesis" (RH) models. The premise of the LNT model is that any dose of radiation above background – and we are all exposed to some degree of background radiation from the many sources mentioned above – can increase the risk of harm with the higher the dose the higher the risk. The RH model proposes that, at very low doses, exposure is actually beneficial and risk of harm only increases beyond a certain dose.

As far as low dose radioactive exposure is concerned, risk relates mainly to the risk of developing some type of cancer, and the two main types of cancer that are thought to be associated with exposure to radioactivity are thyroid cancers and leukemia. The increased risk of developing the former, i.e., thyroid cancer, is thought to be due to the one particular source of radioactivity – iodine 131. The reason for this is because iodine- whether radioactive or non-radioactive – can be taken up specifically by the thyroid gland and used to make the thyroid hormones, T2, T3 and T4. So, if iodine 131 finds its way into our bodies, it is absorbed and therefore concentrated by the thyroid gland. If it is radioactive, then the radiation is confined to the thyroid rather than distributed throughout the body and so it is much more likely to increase the risk of cancer. We know from the accident at Chernobyl that children in particular have a higher risk of developing thyroid cancer when exposed to iodine 131. In addition, it has also been suggested that there is an even higher risk in those children who are iodine deficient.

Because iodine 131 has a short half life (8 days) changes in the thyroid gland that eventually lead to thyroid cancer occur very early on but may take years to develop into cancer. On the other hand, because other radioisotopes such as, for example, cesium 137 and strontium 90 have much longer half lives, if they are taken up into our bodies, and are not excreted quickly, our organs are exposed to ionizing radiation for much longer periods which can increase the risk of cancer. However, as mentioned earlier, our organs vary greatly in their sensitivity to ionizing radiation, so whether cancer develops also depends on where the radioisotope lodges. And, overriding all of this is the question of dose because there does appear to be a relationship between dose and the risk of cancer.

What Do We Know of the Effects of Small Amounts of Radioactivity Exposure on Health?

Much of the research into the radiation exposure on our health is based on epidemiological studies carried out on populations exposed to greater than normal amounts of radioactivity. These studies include survivors of the atomic bomb in Hiroshima and Nagasaki, and accidental exposure as occurred in Chernobyl, Three Mile Island and Fukushima.

Atomic Bomb Survivors

The most comprehensive studies on the effects of exposure to small amounts of radioactivity were those carried out on 94,000 atomic bomb survivors who were present within 10 km from the site of the explosion in Hiroshima and Nagasaki and from 26,000 who were absent at the time of the explosion but later returned to live in these cities. While the amount of radiation exposure within the hypocentre (where the explosion occurred) was of course very high, exposure diminished progressively so that, in Hiroshima the level of exposure dropped from 7 Gy at 1 km to 13 mGy or around one five hundredth, at 10 km from from the hypocentre. With this information, the approximate exposure levels were determined for each of the survivors at the time of the explosion and for a number of years afterwards. At the same time, their health was monitored over a number of decades [15]. The researchers found that the risk of the various forms of leukemia were increased, especially for those survivors who were around 10 years at the time of the exposure and received 1 Gy or more of radiation. The risk of the so-called "solid cancers" which included cancers affecting organs such as breast, lung, brain, etc. (as opposed to leukemia and other blood cancers) was also higher in survivors but was also lower if the exposure occurred in their adult years. Another conclusion was that the risk of cancers was related to the dose of radiation received, higher doses increasing the risk. While there were suggestions that at very low doses, that is less than 10 mGy, there were slightly elevated risks, the findings were not unequivocal.

There has been some criticism of the linear no-threshold model, that is that any amount above normal background radiation can increase the risk. It has been argued that the atomic bomb survivor studies do not support this view and, moreover, regulations that have been developed based on this particular paradigm have resulted in greatly increased economic costs and have restricted the development of the nuclear industry [16].

Accidents

We know that the levels of background radiation we are all exposed to is rarely in excess of 1–10 millisieverts per year which is a lot less than the exposure of most of the atomic bomb survivors. After the accident in Chernobyl, the levels of exposure, particularly for the emergency workers and clean up workers (termed "liquidators") were significantly in excess of this. Many of the 600 emergency workers developed radiation sickness and 28 died [17]. Residents of the most affected areas also received higher than normal doses in the short term but the levels diminished with time. The liquidators showed an increased incidence of thyroid cancer and leukemia [18, 19]. As far as thyroid cancer is concerned, a World Health Organisation report has concluded that in the few years after the accident, the numbers of children with thyroid cancer increased dramatically from a pre-accident incidence of less than one per million to about 40 per million. However, a similar effect was not observed in adults. It is thought that children are particularly sensitive to iodine 131 but these figures clearly show that most children who were exposed to radiation did not develop thyroid cancer. It is believed that some of the radioactive iodine was deposited on pastures and was present in the milk that was produced by cows and later fed to children. There have been suggestions as well that there may be a small increase in breast cancer but more studies are required to confirm this [20].

It is extremely difficult to come up with the minium dose of radiation required to increase the risk of cancer because of all of the variables listed above. The Chernobyl studies suggest – but do not prove – that doses in excess of 100 mSv are required. This figure is many times greater than our normal background exposure. However, it does indicate that small doses of radiation can increase the risk.

There is one further, and perhaps more subtle effect, and that is on the sex ratio i.e., the ratio of males to females. In most countries, there is slight excess of male births, the ratio of male to female births averaging 1.04–1.06. Analysis of the numbers of births after the bombing of Hiroshima, the accident at Chernobyl, the fire at the Windscale reactor, and near nuclear facilities in different parts of Europe, has confirmed a slight increase in the male birth rate particularly in higher exposure areas [21].

There are a number of caveats to all of this. Firstly, it is apparent that, in relation to our indivual risk of developing cancer from radiation exposure, we are not all the same. There is now evidence that, because of genetic factors we do not fully understand, some of us have a greater susceptibility to develop cancer from exposure to radiation. Secondly, we know that, for some cancers, children may be more vulnerable than adults. Thirdly, the marked increase in thyroid cancer as a consequence of the Chernobyl accident has taught us that some radioisotopes are more carcinogenic than others. Finally, Chernobyl has also shown us that how we are exposed to radiation – i.e., whether full body exposure, inhalation or ingestion in food, can also contribute to risk.

In 2011, considerable amounts of radioactivity were released from the nuclear reactors in Fukushima, Japan, as a consequence of an earthquake and a subsequent tsunami. The radioactive plumes contained both cesium 134 and 137, and iodine 131. Analysis of the food and water from the regions not far from Fukushima indicated that the levels of radioactivity in food, measured as cesium 137, even here did not exceed one millisievert. At the time of writing, there was limited information on the presence of radioactive iodine in food. There was a report showing that iodine 131 was present in the breast milk of women living in a city 140 km from the Fukushima reactor site almost 6 weeks after the explosion. It is not known where this iodine came from but there are indications at least some must have come from food and water.

There are a number of grounds for concern. Firstly, if iodine 131, which more than likely came from the Fukushima reactor, travelled the 140 km distance and was taken up into breast milk then those women that were closer to the reactor site were likely to have received a bigger dose. Also, it is not clear whether measurements of breast milk were taken nearer to the time of the accident, as well as in women who lived closer to Fukushima. This is the only way we will be able to ascertain the consequences of exposure to radioactive iodine. Finally, because of the sensitivity of children to iodine 131 exposure, only time will tell whether some children who lived closer to the reactor sites will develop thyroid cancer.

Does Radioactivity Exposure Increase the Risk of Other Diseases?

Cataracts are opaque deposits in the lens of the eye and increase in frequency with age. Studies on atom bomb and Chernobyl survivors have indicated that radiation exposure can increase the risk of cataracts [22]. It was thought formerly that large doses of radiation were required but there are now indications that smaller doses, perhaps less than 0.8 sievert, can cause cataracts. Cataract formation seems to be related to external rather than internal radiation exposure. There are suggestions that the risk of cardiovascular disease is increased but, again, the data available are not convincing [17].

What Do We Know About the Effects of Non-ionising Radiation on Our Health?

As mentioned earlier, non-ionising radiation includes waves of varying frequency and intensity produced from electric and magnetic fields. Computers, televisions, radios, mobile phones, wifi, overhead wires, magnetic resonance imaging machines, and microwave ovens are some of the modern technologies that use/produce non-ionising radiation. All of us are exposed to these waves. While they do not produce the same sorts of drastic effects of some forms of radioactivity, there are concerns that they may not be as inert as had been previously thought.

Microwave Syndrome

There have been reports of that some individuals exposed to relatively low intensity electromagnetic fields may have what has been termed the "microwave syndrome" or electromagnetic hypersensitivity [23]. This condition is said to be characterized by headaches, fatigue, dizziness, poor concentration, and poor memory and only affects some people who may be particularly sensitive to the radiation [23]. However, there is some research that questions whether this syndrome actually exists [24]. According to this view, symptoms experienced are due to what has been termed the "nocebo" effect – that is, the belief that microwaves are harmful is sufficient in itself to induce symptoms in the person exposed. The jury is therefore still out on whether electromagnetic hypersensitivity exists. It is clear therefore that further research is needed.

Magnetic Resonance Imaging (MRI)

While the effects of exposure to low intensity electromagnetic radiation are controversial, there is some evidence that continuing exposure to much higher intensity radiation, as may occur in the workplace, may not be negligible. Thus, workplace exposure to the high intensity electromagnetic fields generated by MRI machines has been reported to cause vertigo in factory workers and nurses [25, 26].

Mobile Phones

Despite the uncertainties raised by the so-called nocebo effects of microwaves, there is some evidence that mobile phones may be harmful. Mobile phones emit and receive electromagnetic energy in the form of radiofrequency waves. As the intensity of this energy is higher closest to the phone, it follows that that those body tissues closest to the phone will receive the greatest amount of energy. The ear, skull, and the brain are therefore the areas of the body that are most exposed to the radiation. While there have been claims that radiation from mobile phones may increase the risk of nausea, Parkinson's disease, headaches, high blood pressure, infertility etc., there is little supportive evidence for this [27]. However, there is some research that seems to show that there may be an association between mobile and cordless phone use and an increased risk of glioma, an aggressive form of brain cancer [28, 29].

WiFi

Despite the fact that wifi is now ubiquitous, there are relatively few studies showing effects on humans. Recent research on rabbits has shown that wifi antennae placed around 25 cm from the heart and generating high intensity fields, induces increases in blood pressure and heart rate [30].

What Does All This Mean?

Humans have been exposed to small amounts of ionizing radiation from a variety of sources throughout the ages. The foods that we eat, and the water that we drink or bathe in, the air that we breathe, all contain small amounts of radioactive substances, as do the fossil fuels such as coal and oil that we use as sources of energy. Similarly, we have been exposed to natural sources of non ionizing radiation via sunlight and electrical storms. The greatest concern relates to the potential effects of ionizing radiation which has the capacity to damage tissues and rheir constituent proteins and DNA. In particular, changes to the latter can increase the risk of cancers and there is evidence for this from survivors of the atomic bomb in Japan and accidents such as those that occurred in Chernobyl. While it is likely that we have developed mechanisms for dealing with small amounts of radioactivity that occur naturally, both the variety and the intensity of ionizing radiation we are exposed to has changed dramatically in the last few decades and whether it is affecting our hrealth remains controversial. According to one school of thought – the no threshold view- any amount of radioactivity has the potential to harm us. The other view is that any risk to our health only occurs when the amount we are exposed to is above a certain level.

We are also being increasingly exposed to non ionizing radiation through the use of new technologies such as the internet, mobile phones, and magnetic resonance imaging machines. While this type of radiation was formerly thought to have minimal effects on our health and wellbeing, there is now some evidence that it may not be completely harmless. Thus there are suggestions that mobile phone use may increase the risk of brain cancer and exposure to relatively high intensity radiation used in magnetic resonance imaging machines over long periods, for example as occurs in an occupational setting, may cause vertigo and nausea in susceptible individuals. As the use of these new technologies spreads from the developed to the developing world, and new technologies are developed, human beings are being exposed to these different forms of non ionizing radiation on a scale that is greater than perhaps at any time in history. The likely impact on our health remains uncertain.

References

1. Bevermann M et al (2012) Occurrence of natural radioactivity in public water supplies in German: (238) U, (234)U, (228)RA, (226)RN, (210)PB, (210)PO and gross alpha activity concentrations. Radiat Prot Dosim 141:72–81. http://www.ncbi.nlm.nih.gov/pubmed/20413420
2. Povinec PP et al (2005) 90Sr, 137CS, and (239,240)Pu concentration surface water time series in the Pacific and Indian Oceans –WOMARS results. J Environ Radiaoct 81:63–87. http://www.ncbi.nlm.nih.gov/pubmed/15748662
3. Kristbjornsdottir A, Rafnsson V (2012) Incidence of cancer among residents of high temperature geothermal areas in Iceland: a census based study 1981–2010. Environ Health 11:73. http://www.ncbi.nlm.nih.gov/pubmed/23025471

4. Barish RJ (1999) In-flight radiation: counseling patients about risk. J Am Board Fam Pract 12:195–199. http://www.ncbi.nlm.nih.gov/pubmed/10395415
5. Technologically-enhanced, naturally-occurring radioactive materials. USA Environmental Protection Agency. http://www.epa.gov/rpdweb00/tenorm
6. USA Environmental Protection Agency. Nuclear submarines. http://www.epa.gov/radtown/submarine.html
7. Beresford NA et al (2007) The transfer of radionuclides from saltmarsh vegetation to sheep tissues and milk. J Environ Radioact 98:36–49. http://www.ncbi.nlm.nih.gov/pubmed/17765368
8. Al-Kharouf SJ et al (2008) Natural radioactivity, dose assessment and uranium uptake by agricultural crops at Khan Al-Zabeeb, Jordan. J Environ Radioact 99:1192–1199. http://www.ncbi.nlm.nih.gov/pubmed/18359539
9. Lenka P et al (2013) Ingestion dose from 238U, 232Th, 226Ra, 40K and 137Cs in cereals, pulses and drinking water to adult population in a high background radiation area, Odisha, India. Radiat Prot Dosim 153:328–333. http://www.ncbi.nlm.nih.gov/pubmed/22802517
10. Nasim-Akhtar S-J, Tufail M (2007) Natural radioactivity intake into wheat grown on fertilized farms in two districts of Pakistan. Radiat Prot Dosim 123:103–112. http://www.ncbi.nlm.nih.gov/pubmed/17185310
11. Boukenfouf W, Boucenna A (2011) The radioactivity measurements in soils and fertilizers using gamma spectrometry technique. J Environ Radioact 102:336–339. http://www.ncbi.nlm.nih.gov/pubmed/21334798
12. Bevermann M et al (2010) Occurrence of natural radioactivity in public water supplies in Germany: (238)U, (234)U, (235)U, (226)RA, (210)PN, (210)PB, (210)PO and gross alpha concentrations. Radiat Prot Dosim 141:72–81. http://www.ncbi.nlm.nih.gov/pubmed/20413420
13. Hatch M et al (2005) The Chernobyl Disaster: Cancer following the accident at the Chernobyl Nuclear Power Plant. Epidemiol Rev 27:55–66. http://epirev.oxfordjournals.org/content/27/1/56.full
14. Seong KM et al (2016) Is the linear no-threshhold dose-response paradigm still necessary for the assessment of health effects of low dose radiation? J Korean Med Sci 31(Suppl 1):S10–S23
15. Ozasa K et al (2012) Studies of the mortality of atomic bomb survivors, Report 14, 1950–2003: an overview of cancer and noncancer diseases. Radiat Res 177:229–243
16. Socol Y, Dobrzynski L (2015) Atomic bomb survivors Life-Span study: insufficient statistical power to select radiation carcinogenesis model. Dose Response 13
17. Hasegawa A et al (2015) Health effects of radiation and other health problmes in the aftermath of nuclear accidents with an emphasis on Fukushima. Lancet 386:479–488
18. Ostroumova E et al (2014) Thyroid cancer incidence in Chernobyl liquidators in Ukraine. Eur J Epidemiol 29:337–342
19. Ivanov VK et al (2012) Leukemia incidence in the Russian cohort of Chernobyl emergency workers. Radiat Environ Biophys 51:143–140
20. Pukkala E et al (2006) Breast cancer in Belarus and Ukraine after the Chernobyl accident. Int J Cancer 119:651–658
21. Scherb H et al (2015) Ionising radiation and the human gender proportion at birth – a concise review of the literature and complementary analyses of historical and recent data. Early Hum Dev 91:841–850
22. Hammer GP et al (2013) Occupational exposure to low doses of ionizing radiation and cataract development: a systematic literature review and perspectives on future studies. Radiat Environ Biophys 52:303–309
23. Carpenter DO (2015) The microwave syndrome or electro-hypersensitivity: historical background. Rev Environ Health 30:217–222
24. Berthelot M (2016) Is electromagnetic hypersensitivity entirely ascribable to nocebo effects? Joint Bone Spine 83:121–123

25. Gorlin A et al (2016) Occupational hazards of exposure to magnetic resonance imaging. Anaesthesiology 123:976–977
26. Schaap K et al (2016) Exposure to MRI-related magnetic fields and vertigo in MRI workers. Occup Environ Med 73:161–166
27. Kim KH et al (2016) The use of cell phone and insight into its potential health impacts. Environ Monit Assess 188:221
28. Morgan LL et al (2015) Mobile phone radiation causes brain tumours and should be classified as a probable human carcinogen (2A) (review). Int J Oncol 46:1865–1871
29. Hardell L et al (2013) Case-control study of the association between malignant brain tumours diagnosed between 2007 and 2009 and mobile and cordless phone use. Int J Oncol 43:1833–1845
30. Saili L et al (2015) Effects of acute exposure to WIFI signals (2.45GHz) on heart rate variability and blood pressure in Albinos rabbit. Environ Toxicol Pharmacol 40:600–605
31. Suzuki K, Yamashita (2012) Low-dose radiation exposure and carcinogenesis. Jpn J Clin Oncol 42:563–568. http://www.ncbi.nlm.nih.gov/pubmed/22641644
32. Ng KH (2003) Non-ionizing radiations – sources, biological effects, emissions, and exposures. In: Proceedings of the international conference on non-ionizing radiation at UNITEN, 20–23 October 2003. http://www.who.int/peh-emf/meetings/archive/en/keynote3ng.pdf
33. Mathew S et al (2012) Natural radioactivity content of soil and indoor air of Chellanam. Radiat Prot Dosim 152:80–83. http://www.ncbi.nlm.nih.gov/pubmed/22951996
34. Ngachin M et al (2008) Radioactivity level and soil radon measurement of a volcanic area in Cameroon. J Environ Radioact 99:1056–1060. http://www.ncbi.nlm.nih.gov/pubmed?term=radon%20AND%20sources%20AND%20cameroon

Chapter 13
E-Waste

The digital revolution of the last few decades of the twentieth century has brought enormous benefits to the lives of human populations in both developed and developing countries. It has transformed our lives in so many ways. Its effects were shown perhaps even more clearly during the coronavirus pandemic when movement was often restricted and ordinary face to face meetings were often replaced by "virtual meetings". Even visits to the doctor were not immune as there was increasing reliance on so-called "telehealth" which was carried out via computer screens. While there are clearly enormous benefits to digital technology there is another side which is perhaps not fully appreciated and that is the technology relies on the use of electrical and electronic equipment which, like many other industrial products made by humans generates what is often referred to as "e-waste".

E-waste is derived from a great variety of equipment and devices including electrical and electronic equipment such as household appliances (e.g., refrigerators, washing machines), consumer equipment (e.g., TVs, phones, DVD players), information technology (e.g., computers, monitors), and transport (trams, trains, and aircraft). The batteries, switches, cathode ray tubes, circuit boards, toner cartridges etc. that are major components of the various devices and equipment may contain a variety of materials such plastics, glass, metals, flame retardants, and refrigerants [1]. Plastics in particular are a major component of e-waste and include polymers such as acrylonitrile butadiene styrene, high impact polystyrene, polycarbonate, polyamide, polypropylene, polyurethane, polyvinylchloride, and epoxies [2, 3].

According to a United Nations report published in 2019 the world produced 44 million tonnes of electrical and electronic waste in 2017 and is on track to generate 120 million tonnes by 2050. According to this same report only 20% is recycled with the rest either being sent to landfill or recycled manually by people who live in undeveloped countries [4].

Newer Forms of E-Waste

Because of concerns about the effects of carbon dioxide emissions from fossil fuels, there has been considerable research into alternative forms of power generation. Electric cars have been developed and are likely to take over from petrol or diesel driven cars. Power generated from solar panels and windmills are gradually replacing coal and oil driven power stations. While there are clear environmental advantages in the development of these alternative forms of energy it cannot be assumed that there are no environmental issues associated with the use of these technologies. For example while electric cars release little carbon dioxide and hydrocarbon breakdown products, significant amounts of particulate matter (PMs mentioned in an earlier chapter) formed from, for example brakes, tyres, and road surfaces, are produced. E-waste is also generated from the various electronic components of electric vehicles, in particular the batteries, often containing lithium and having a life of 3–5 years [5].

The increasing use of solar panels, which are an integral part of newer forms of energy production, also has environmental implications, although they mostly have a long service life, perhaps lasting up to 25 years. A typical photovoltaic cell module is comprised of frames, covers, encapsulation layers, solar cell, cables, and conductors and its various components include glass, aluminium, adhesives, polyvinyl fluoride, copper, silver, and lead [6].

Another form of energy is from wind power. The blades are often made of a polymer composite, often epoxy and polyester, and reinforced with glass and carbon fibre and thermoplastic matrices [7]). They are difficult to recycle so may be sent to landfill. It is not an insignificant problem because it is estimated that there will be around 800,000 tonnes of blade waste by 2050 [8]. The NdFeB magnets in wind turbine generators have a limited life and are recycled to obtain the rare earth metals [9, 10]. Other potential e-waste from this source would include batteries used for storage of electrical energy and a variety of other electronic components.

Environmental Release of E-Waste Chemicals

Whichever way the waste is treated, there seems little doubt that many of its chemical components are released into the environment either through landfills, recycling, or burning. For example, the chemicals released from the breakdown products of plastics include a great variety of different substances such as pigments, stabilisers, antioxidants, plasticisers, lubricants, and flame retardants. It is worth noting that each additive class itself may include a number of different chemicals. For example, at least six plasticisers, seven stabilisers, and five antioxidants have been detected in plastic debris [11]. In addition to the plastics, e-waste contains a variety of metals. Recent studies have confirmed the presence of up to 56 elements in a variety of electronic devices and up to 48 metals in computer hard drives. The metals include lead, cadmium, chromium, arsenic, manganese, mercury, lithium, nickel, and copper [3]. Included in these metals up to 14 elements commonly referred to as rare earth elements. These include elements such as yttrium, lanthanum, cerium, gadolinium etc. Further treatment of e-waste may add another layer of complexity to the chemical composition of e-waste. For example the burning of e-waste can generate a number of potentially toxic dioxin-like chemicals, polycyclic hydrocarbons and other substances [12, 13]. There are as well other more subtle changes. For example, the presence of some of the e-waste pollutants can lead to changes in the composition of the microbial communities that live in the soil [14, 15]. Whether this can affect the future fertility of the soil, or whether some of the products of these different microbial communities can enter the food chain, is really not known.

What Are the Effects of E-Waste Components on Our Health?

In order to produce any adverse effects the chemical components of e-waste released into the environment must be taken up by animals and humans. As some of the chemicals present in e-waste sites, such as for example, the flame retardants, have been found in the tissues of birds, insects, reptiles, and fish it is clear that some at least they are taken up by wildlife [16]. There is also no doubt that they some pollutants from e-waste sites find their way into soil, food, water, and air and from there at least some of the e-waste components are also taken up by humans into their tissues and body fluids [17–19]. For example, PBDEs often used as flame retardants, and polycyclic hydrocarbons, have been found in elevated amounts in breast milk and a variety of tissues and body fluids in e-waste recycling areas and in e-waste recycling workers [19, 20]. Rare earth elements such lanthanum and cerium have been shown to be present in increased amounts in the blood of residents in an e-waste recycling area [21]. Exposure to toxic metals such as lead and mercury have also been found to be greater at or near e-waste sites [22–24].

As at least some of the chemicals are clearly taken up into our tissues, is there any evidence that they are harmful? One way of assessing possible health related effects is through examining the health of individuals who have the highest exposure to e-waste. These are mostly those who are exposed to higher levels of e-waste chemicals through their working environment. Most of these are workers in the formal and informal recycling industries. The latter is mostly carried out in developing countries where there may be deficiencies in environmental management as compared to the formal recycling sector which is mostly larger and where there is provision for better health and environmental controls. Another way of assessing the effects of e-waste chemicals is to examine the health of individuals (i.e., non workers) who live near an existing or former e-waste site.

In a study carried out on individuals from Taizhou, one of the largest e-wasted recycling areas in China and which was closed in 2015, nine metals, including chromium, arsenic, cobalt, nickel, tin, mercury, lanthanum, and cerium, were found in elevated amounts in the blood as compared to a non recycling reference area. The levels of three important hormones – the corticotropin releasing hormone, adreno-corticotropin hormone and cortisol were found be increased in residents from the e-waste recycling area and it was speculated that these hormonal changes related to the increased uptake of pollutants, and particularly lead. These hormones are a part of the so-called hypothalamus-pituitary-adrenal gland axis which regulate different critical activities in the body [25]. In addition, the exposed group showed evidence of an increase in oxidative stress which is a measure of how well the body removes potentially harmful free radicals. In another study, increased lead exposure was believed to have resulted in the hearing loss observed in 3–7 year old children who lived near a e-waste recycling area. The higher incidence of hearing loss in exposed children was thought to be due to effects of lead, and perhaps other pollutants, on the developing auditory system in the children [26]. There are as well indications that maternal exposure to lead from e-waste sites can result in changes in some of the brain cell DNA in newborn babies [27]. It is thought as well that lead from e-waste sites may increase the risk of dental caries through its effect on anti-inflammatory processes in the mouth [28]. The importance of lead in e-waste is clear from research showing that increases in blood, urine and hair levels of lead are thought to act as "biomarkers" of exposure i.e., increased body levels of lead are a sign of excessive exposure [29]. It is important to note that another group of metals, the rare earths, in particular cerium and lanthanum, are found at e-waste dismantling sites, and have also been implicated in an increase in the levels of the thyroid stimulating hormone which in turn regulates the amount of thyroid hormone produced by the thyroid gland [21].

In addition to exposure to metals present in e-waste sites a variety of other pollutants have been detected in human tissues and may have harmful effects. For example, halogenated and organophosphate flame retardants which have been added to reduce the risk of fire in electronic products, have also been detected in a variety of human tissues including blood, hair, breast milk, nails, abdominal fat, and urine [30]. The production and use of some of these chemicals, for example the polychlorinated biphenyls and the polybrominated diphenyl ethers, was discontinued a number of years ago or their use has been restricted but because of their environmental stability, they are still detected at some e-waste sites and are taken up into human tissues and body [31].There are indications that some of these pollutants also increase the levels of oxidative stress and may interfere with the action of thyroid hormones [37].

Plastics, and in particular their breakdown products, microplastics, represent yet another source of pollution from e-waste sites and have been mentioned elsewhere in this book. Many different types of microplastic have been detected in the soil from e-waste sites. Of particular concern is the presence of different metals adsorbed to the plastic, including the more toxic metals such as lead and cadmium, which are known to increase the risk of disease [32].

There have been limited studies on the potentially harmful effects of the rapidly expanding solar energy industry with its with its solar panels and photovoltaic cells. Even though the panels are considered relatively inert, over time and with exposure to heat and water, crushed or broken modules almost certainly break down over time releasing some of the potentially toxic constituent metals such as lead and cadmium into the atmosphere, soil and water [33, 34]. In addition, back sheet components of the photovoltaic cells often include plastics containing the highly reactive element fluorine (polyvinylidene fluoride and polyvinyl fluoride). Incineration of these back sheets may release a variety of fluorine-containing substances including toxic dioxins [35].

There are indications that some of the materials used in the manufacture of the blades for wind power (i.e., epoxy resins) may increase the risk of skin and lung irritation at least in an occupational setting although there is no evidence as yet that the recycling of blades produces the same effect [36]. NdFeB magnets which are often used in wind turbines contain mixtures of metals including the rare earths neodymium, praseodymium, and dysprosium, which have been mentioned earlier in relation to possible effects on the function of the thyroid gland [9].

What Does This Mean?

The 2020/2021 coronavirus pandemic showed very clearly how the digital world has shaped our lives and changed the way we live. At the same time there has been increasing pressure on governments around the world to reduce our reliance on fossil fuels because of their probable impact on global warming and this in turn has led to the development of alternative sources of energy such as wind and solar power and the gradual substitution of electric cars for fossil fuel driven motor vehicles. While the use of alternative sources of energy will certainly reduce the release of carbon dioxide as well as other potentially harmful products from fossil fuels, it is perhaps not fully appreciated that there are environmental costs to this changing world. The new technologies also generate waste which either through recycling, landfill, or burning also release chemical pollutants into the environment. And like other forms of waste, chemicals in e-waste find their way into the atmosphere, water, soil and food, and from there into wildlife and even into human tissues. There is already evidence that some chemicals in e-waste can interfere with basic functions in animals and in humans. As the amounts, and diversity of electronic products increases, so too will the amounts and variety of e-waste and its chemical components. There is no doubt that new chemicals will be developed and used to fulfil specific needs. Unfortunately, If the past is any guide to what is likely to happen in the future, many of these chemicals will be incorporated into new electronic products without a proper understanding of their likely impact on the environment and our health and wellbeing.

References

1. Li W, Achal V (2020) Environmental and health impacts due to e-waste disposal in China – a review. Rev Sci Total Environ 737:139745
2. Das P et al (2020) Value-added products from thermochemical treatments of contaminated e-waste plastics. Chemosphere 269:129409
3. Buechler DT et al (2020) Comprehensive elemental analysis of consumer electronic devices: rare earth, precious, and critical elements. Waste Manag 103:67–75
4. UN Report (24 January 2019) Time to seize opportunity, tackle challenge of e-waste. https://www.unenvironment.org/news-and-stories/press-release/un-report-time-seize-opportunity-tackle-challenge-e-waste
5. Zhao Q et al (2020) Recovery and regeneration of spent lithium-ion batteries from new energy vehicles. Front Chem 8:807
6. Xu Y et al (2018) Global status of recycling waste solar panels. A review. Waste Manag 75:450–458
7. Mishnaevsky L et al (2017) Materials for wind turbine blades. An overview. Rev Mater (Basel) 10:1285
8. Liu P, Barlow CY (2017) Wind turbine blade waste in 2050. Waste Mang 62:229–240
9. Kumari A et al (2018) Recovery of rare earths from spent NdFeB magnets of wind turbines: leaching and kinetic aspects. Waste Manag 75:486–498
10. Smith BJ et al (2018) Costs, substitution, and material use: the case of rare earth magnets. Environ Sci Technol 52:3803–3811
11. Rani M et al (2015) Quantitative analysis of additives in plastic marine debris and its new products. Arch Environ Contam Toxicol 69:352–366
12. Dai Q et al (2020) Severe dioxin-like compound (DLC) contamination in e-waste recycling areas: an under-recognised threat to local health. Environ Int 139:105731
13. Nishimura C et al (2017) Occurrence, profiles, and toxic equivalents of chlorinated and brominated polycyclic aromatic hydrocarbons in E-waste open burning soils. Environ Pollut 225
14. Wu Q et al (2019) Vertical profile of soil/sediment pollution and microbial community change by e-waste recycling operation. Sci Total Environ 669:1001–1010
15. Wu Z et al (2019) Effects of soil properties, heavy metals, and PBDEs on microbial community of e-waste contaminated soil. Ecotoxicol Environ Saf 180:705–714
16. Liu Y et al (2018) Halogenated organic pollutants in aquatic, amphibious, and terrestrial organisms from and e-waste site. Habitat-dependent accumulation and maternal transfer in watersnake. Environ Pollut 241:1063–1070
17. Cai K et al (2020) Human exposure to PBDEs in e-waste area: a review. Environ Pollut 267:115634
18. Wu Q et al (2019) Trace metals in e-waste lead to serious health risk through consumption of rice growing near an abandoned e-waste recycling site: comparisons with PBDEs and AHFRs. Environ Pollut 247:46–454
19. Asamoah A et al (2019) PAHs contamination levels in the breast milk of Ghanian women from an e-waste recycling site and a residential area. Sci Total Environ 666:347–354
20. Meng T et al (2021) Global distribution and trends of polybrominated diphenyl ethers in human blood and breast milk: a quantitative meta-analysis of studies published in the period 2000–2019
21. Guo C et al (2020) Rare earth elements exposure and the alteration of hormones in the hypothalamic-pituitary-thyroid (HPT) axis of the residents in an e-waste site: a cross sectional study. Chemosphere 252:126488
22. Zeng X et al (2020) E-waste lead exposure and children's health in China. Sci Total Environ 734:139286

23. Tang W et al (2015) Mercury levels and estimated daily intakes for children and adults from an electronic waste recycling area in Taizhou, China: key role of rice and fish consumption. J Environ Sci (China) 34:107–115

24. Wu Q et al (2019) Trace metals in e-waste lead to serious health risk through consumption of rice growing near an abandoned e-waste recycling site. Comparison with PBDEs and AHFRs. Environ Pollut 247:46–54

25. Li Z et al (2020) The sustaining effects of e-waste-related metal exposure on hypothalamus-pituitary-adrenal axis reactivity and oxidative stress. Sci Total Environ 739:139964

26. Liu Y et al (2018) Hearing loss in children with e-waste lead and cadmium exposure. Sci Total Environ 624:621–627

27. Zeng Z et al (2019) Differential DNA methylation in newborns with maternal exposure to heavy metals from an e-waste recycling area. Environ Res 171:536–545

28. Hou R et al (2020) Elevated levels of lead exposure and impact on the anti-inflammatory ability of oral sialic acid among preschool children in e-waste areas. Sci Total Environ 699:134380

29. Arain AL, Neitzel RL (2019) A review of biomarkers used in assessing human exposure to metals from e-waste. Rev Int J Environ Res Public Health 16:1802

30. Ma Y et al (2021) Human exposure to halogenated and organophosphate flame retardants through informal e-waste handling activities. A critical review. Rev Environ Pollut 268 (Pt A):115727

31. Meng HJ et al (2020) Levels and sources of PBDEs and PCBs in human nails from e-waste, urban, and rural areas in South China. Environ Sci Process Impacts 22:1710–1717

32. Chai B et al (2020) Soil microplastic pollution in an e-waste dismantling zone in China. Waste Manag 118:291–301

33. Nain P, Kumar A (2020) Ecological and human health risk assessment of metals leached from end-of-life solar photovoltaics

34. Kwak J et al (2020) Potential environmental risk of solar cells: current knowledge and future challenges. J Hazard Mat 392:122297

35. Danz P et al (2019) Experimental study of fluorine release from photovoltaic backsheet materials containing PVF and PVDF during pyrolysis and incineration in a technical lab-scale reactor at various temperatures. Toxics 7:47

36. Freiberg A et al (2018) Health effects of wind turbines in working environments – a scoping review. Rev Scand J Work Environ Health 44:351–369

37. Curtis SW et al (2019) Thyroid hormone levels associate with exposure to polychlorinated biphenyls and polybrominated biphenyls in adults exposed as children. Env Health 18:75

Chapter 14
Do Environmental Chemicals Cause Disease and, If So, How?

One of my father-in-law's favourite sayings was "Old age is no asset". And, as far our health is concerned, this is quite clearly true because our bodies do deteriorate with time and the likelihood of degenerative diseases like cancer and heart disease, do increase with age. However, while age is clearly a very important factor, it is not the only one. We know, for example, that cancer, heart disease and diabetes do affect children and young adults so the processes that lead to degenerative disease probably start very early in life but our bodies are mostly better able to repair any damage before it leads to disease. Why some children develop these sorts of diseases is not really understood. It may be that the type and magnitude of the genetic predisposition overwhelms the repair processes in these children. Whatever the reason, degenerative disease is relatively uncommon in the young and is mostly associated with getting older. It is important to emphasise that while cancer and heart disease are the major killers, there are many other degenerative diseases and these include chronic fatigue syndrome or fibromyalgia, type 2 diabetes, motor neurone disease, multiple sclerosis, asthma, most cancers, autoimmune (the body's immune own immune system attacking various organs) diseases such as type 1 or the childhood form of diabetes, lupus and scleroderma etc., for which there is no known cause.

Pessimists argue that there is nothing we can about it and we just have to accept getting sick as we age. That may be true but the issue is not that we will not eventually succumb to the ravages of time, because sooner or later we all will, but whether we can delay the onset of degenerative disease as long as possible. At present it seems that there is an acceptance that we really can't do much about it and, anyhow, with the use of drugs, most of us can live fairly comfortably into our eighties. However, while drugs do help to keep us alive a lot longer, and new and better drugs are being developed all of the time, they are not without their problems. They are expensive, both for the patient and the community, can cause serious side effects in some people and they do not work for everyone. Wouldn't it be better if we could live relatively disease free for a decade or more without the use of drugs? The question is – is it possible?

© The Author(s), under exclusive license to Springer Nature Switzerland AG 2021   189
A. Poulos, *The Secret Life of Chemicals*,
https://doi.org/10.1007/978-3-030-80338-4_14

This question is closely linked to yet another question – i.e., what is the natural life span of homo sapiens? We know that a thousand or so years ago most people did not live much beyond fifty. People who did survive childhood often died prematurely from a variety of causes such as wars, infections, and malnutrition. It is still true today to a certain extent in parts of the third world, and even in certain groups in developed countries i.e., Australian aborigines. In third world countries the life span is shorter because of diseases such as malaria, TB and, more recently AIDS that have largely been eradicated or controlled in Western countries, while indigenous peoples who live in developed countries, such as the American Indians and the Australian aborigines have higher incidences of Western diseases such as diabetes and heart disease. What this tells us is that the natural life span is at least the 80 plus which is found in most developed countries. There are reasons for believing that our potential lifespans may be even greater than this because the proportions of centenarians in our midst is increasing and, moreover, there are certain regions of the world, for example parts of Sardinia and in Okinawa, where the proportion is even higher. But while survival to a hundred is impressive, the more important question is whether the centenarians in these regions are still able to have a normal life.

This question was addressed by Willcox et [1] who carried out a study into the inhabitants of Okinawa, a group of Japanese islands to the south west of Kyushu. In his book, "The Okinawa Way", he speculates that there are many factors that contribute to the longevity of Okinawans. These factors include diet, lifestyle, social support, exercise, and and meditation. There is no mention of chemicals in the environment although as tourism, agriculture and fishing are the main activities on the islands, at least up to relatively recently, Okinawans are clearly not as exposed to the great diversity of industrial pollutants as the inhabitants of the major Japanese cities such as Tokyo. In addition, there is increasing evidence that genetic factors may also contribute to healthy aging and longevity.

In most cases, it is true that doctors and other health professionals treat our symptoms without knowing the cause of our disease. So, for example, if we have had a heart attack we can have stents put in, or have a bypass, but our doctor may not really know why the plaque, which was perhaps the immediate cause of our heart attack, built up in our blood vessels. What they do know, however, is that there are certain factors, such as high blood cholesterol, or triglycerides which could have played some part in the process so we may be advised to change our diets and take statins or enriched fish oils to lower our blood cholesterol and triglycerides respectively thereby reducing our risk. The problem with this is that it does not really address the true cause of plaque – that is why did the plaque build up in the first place? Indeed, almost 50% of people who have had heart attacks do not have high cholesterol levels, so cholesterol is clearly not the answer. It is even more complicated in the case of cancer because there are so many different cancers. There are a few risk factors that we know about, but these are for specific cancers, such as the BRCA1 and BRCA2 genes in breast cancer, and again, these risk factors do not tell us how our cancer developed. And cancer and heart disease are not the only degenerative diseases, and nor are they the only diseases for which we know little about the actual cause. Even less is known about the causes of conditions such as chronic fatigue syndrome or fibromyalgia, type 2 diabetes, motor neurone disease, multiple sclerosis, asthma, most cancers, autoimmune (the body's immune own immune system attacking various organs) diseases such as type 1 or the childhood form of diabetes, lupus and scleroderma. From genetic studies, and particularly studies involving twins, we do know that disease is generally a response of our bodies, and more specifically, to environmental factors. For example, if have a bacterial or viral infection, specific genes are activated leading to the production of special proteins in different parts of our bodies, particularly the immune system, and these proteins are involved in the many different mechanisms that take part in combating infections. Specific genes are also activated in response to environmental factors, including chemicals, although these genes are different from those involved in bacterial or viral infections.

What Is the Evidence that Chemicals Cause Disease?

From studies with animals we know that certain chemicals can cause diseases in animals. For example, there is a very great number of chemicals that can cause cancer (the IARC list of carcinogens referred to earlier), diabetes can be produced simply by administering chemicals such as streptzotocin in rats, Parkinson's s disease can be induced with MPTP, collagen and carrageenan can induce arthritis, autoimmune conditions such as lupus can be induced by pristane (a component of mineral oil), there are chemicals that can induce abnormalities of the heart muscle, asthma and skin diseases can be produced at will with different chemicals, and there are a variety of substances that can produce birth defects in animals.

As far as humans are concerned, there is little doubt that chemicals do cause disease, the type of disease probably depending on the type of chemical, the amount and duration of exposure, and probably many other variables such as sex, age, diet, lifestyle etc. The adverse effects of some pharmaceuticals, often when either adminstered or taken in relatively small amounts, is convincing evidence that chemicals in relatively small amounts can cause disease [1]. The teratogenic effects of thalidomide, antiepileptic drugs such as valproic acid, and even aspirin are good examples of the ability of chemicals to induce changes even in the developing foetus.

As will be discussed in the next chapter, it seems fairly clear that there are some people who appear to be more susceptible. However, how do amounts of environmental chemicals that are an order of magnitude lower than amounts required to cause harm in animals contribute to diseases such as cancer, diabetes, and heart disease in people? There are a number of possible explanations but, before we examine these, it is worth looking at what happens to those chemicals that enter our bodies.

How Do Environmental Chemicals Enter Our Bodies and What Happens to Them?

For an environmental chemical to produce an effect on our health it has to be taken up into our body. As discussed a number of times throughout this book, there are really three routes of entry – via the mouth, the skin, or the lungs. More recent evidence indicates that there is also a fourth – the nasal mucosa- the lining within the nasal cavity, The latter, through its connection with the olfactory lobe, the part of the brain that involves the sense of smell, is now believed to permit the entry of certain chemicals directly into the brain. Food and water are probably the main source, and environmenal chemicals taken up in this way include the plasticisers, pesticides, arsenic, cadmium, the by products of water chlorination, and the many different industrial chemicals such as PCBs, PBDEs, dioxins, mercury, fluorochemicals, and no doubt a great number of other substances, many of which remain undetected at present but no doubt will be in the future. Most then travel via the blood to the liver where they may be changed chemically into forms that can more readily be excreted by the kidney.

The Role of the Liver and Kidney

The chemical components of food, contaminants present in water, alcoholic drinks, drugs, and chemicals absorbed through the skin, the lungs, or even the brain eventually find their way via the blood into the liver. One of the major functions of the liver is to change the absorbed chemicals into less toxic, and often more water soluble forms, so that they can be excreted by the kidney. There are special proteins in the liver that are involved in this process. As there are many different chemicals, our bodies have to detoxify and then eliminate them via the kidneys, so it is not surprising that there many different proteins involved, each capable of reacting with a particular type of chemical structure or to use the scientific term "specificity". One of the best known is cytochrome P450. It is interesting to note that that there are dozens of genes, producing dozens of different cytochrome P450 proteins, each with a unique specificity, It is even more complex than this because there are many other proteins involved in detoxification. These detoxification processes are particularly important in the elimination of drugs.

While most enter the liver, smaller amounts, which can vary according to the chemical structure, can travel directly to other organs including the pancreas, lungs, brain and even the eye. Travel to the brain is more problematic because of the blood-brain barrier, a type of filter that can block the uptake of chemicals that may damage the brain and mentioned earlier. However, despite this, there is no doubt that some pollutants get into the brain, including lead, cadmium, and organochlorine pesticides. Indeed, some pollutants even manage to find their way into quite specific parts of the brain such as the substantia nigra, the part of the brain that is impaired in Parkinson's disease [1].

There appear to be other mechanisms for getting rid of chemicals that may be harmful because the fat stores (referred to as adipose tissue), hair and even nails, have been reported to contain some environmental chemicals. Storage in these sites may be a way of getting potentially toxic chemicals away from delicate organs.

The fact that many environmental chemicals are taken up into our bodies is beyond dispute. Even fetal tissues and breast milk have been shown to contain a variety of pollutants that no doubt reflect maternal exposure, through food, water, or air. But, given that they are present in different parts of the body, are they completely inert? Or can they disrupt one or more of the myriad of delicate processes that take place in the different cells in our bodies? Let us take a look at possible mechanisms, based on what we know at the present time, that may help us explain the role of environmental toxins in the development of degenerative diseases.

Do Environmental Chemicals Cause Disease?

Certainly in large doses chemicals like PCBs, organochlorine and organophosphorus pesticides, lead, arsenic, cadmium, and mercury, to name just a few, are known to be toxic. In doses that are not small enough to kill immediately but in amounts greater than that normally found in our food or in the environment, we know that they can make people and animals sick. Good examples, and mentioned earlier in this book, include the PCBs in animal feeds, mercury in grain and fish, arsenic in milk powder, lead in toys and pottery, pesticide residues in food, to name just a few. We know, too, that exposure to small amounts of certain chemicals, not enough to cause disease in a single or even multiple doses, can be harmful if the exposure is chronic occurring over many years. Examples include those diseases that develop from arsenic in the drinking water, occupational exposure such as mesothelioma, silicosis, and black lung disease in mining, lung and skin diseases and perhaps certain forms of blindess and Parkinson's disease in farmers and pesticide applicators, white blood cell cancers in workers exposed to benzene, and lung disease in workers exposed to beryllium and diacetyl. The case for the exposure to air pollution, in particular the PMs in air contributing to cardiovascular disease, and exposure to lead in the air and soil contributing to impaired brain function in children, is also quite strong. However, the much more difficult question to answer, and one that has been raised repeatedly in this book, is – can the smaller amounts we are all exposed to outside the work environment for example the traces of pesticides, heavy metals, and plasticisers in our food, and the pollutants in our air and water make us sick? And, if they do make us sick, do we have any idea which diseases are caused by exposure to pollutants? As they say, the jury is still out even though there is enough information now available to answer the question in the affirmative.

Are There Parts of Our Bodies that Are More Vulnerable to the Effects of Environmental Chemicals?

The human body is an incredible structure. On the surface, at least at a non-microscopic level, it appears relatively simple and yet, at a microscopic level, the complexity inherent even in a single cell of any tissue, is truly overwhelming. And yet, examination of a cell under a microscope tells you nothing of the even more astonishing complexity. Apart from the myriad of coordinated biochemical processes that take place every day within each cell, most would be surprised to learn that individual cells in an organ such as the liver or brain are in communication with their neighbours. Furthermore, each organ does not exist in splendid isolation but, through the bodies' transport and communication system, the blood vessels and the nervous system, is in contact with other organs. As our natural state is health, most of the time everything is working well. Unlike other complex structures, if there is a glitch, for example a bacterial infection that threatens to damage body tissues, the immune system steps in, removes the threat, and repairs any damage that may have occurred. But, despite our body's capacity to repair itself, it is subject to a gradual deterioration over time. Scientists have identified many of the factors that appear to hasten the deterioration. Many, if not most, are environmental factors and include the different types of radiation, diet, infective agents, and different types of chemical exposure. These environmental factors produce their effects through their ability to damage certain critical, but vulnerable, processes or structures within the cells of the many organs of our bodies. Two imporant cellular components that are known to be affected by pollutants include the mitochondria and our genes. Chemicals are also known to affect the function of the immune system. How pollutants may produce their effects on these cellular components, and on the the immune system, will be discussed in the following chapters.

The elimination of chemicals that are potentially toxic to the body is of fundamental importance and it is therefore not surprising that there are scores of genes involved. Even in paleolithic times elimination of toxic materials present in the environment was important but, with increased globalisation, industrialisation, as well as rapid population growth, there has been an almost exponential increase in the number and variety of chemicals released into the environment. Of course even in paleolithic times there were chemicals released or generated from volcanic eruptions, forest fires, solar radiation, and lightning, and there was exposure from food and water as there is today, but the number and variety of chemicals we are now exposed to has almost certainly increased greatly. What we really do not know is whether our livers and kidneys can cope with the thousands of new chemicals e.g., PCBs, PBDEs, fluorocarbons etc. released into the environment.

What Does All This Mean?

Exposure to environmental chemicals can occur in many ways, including our food, water, and the air we breathe. The chemicals we are exposed to can enter our bodies via mouth, skin, and lungs. Entry can also occur via the nasal mucosa and, in this case, chemicals may enter the brain directly thereby bypassing the blood-brain barrier, There is little doubt that exposure to certain chemicals can cause disease and even death. Exposure to repeated doses of small, non toxic, amounts i.e., chronic exposure, of certain chemicals may also cause disease depending on the dose, the duration of exposure, and the susceptibility of the individual. It is known that the function of at least two important cellular components are susceptible to chemicals, in particular the mitochondria, the power generating parts of cells, and our genes. Chemicals are also known to affect the function of the immune system. In the following chapters, we will look at ways in which chemical pollutants may interfere with the function of mitochondria, our genes, and our immune systems, thereby increasing the risk of disease.

References

1. Guveli BT et al (2017) Teratogenicity of antiepilectic drugs. Clin Psychopharmacol Neurosci 15:19–27

Chapter 15
Genetic Variability and the Risk of Disease- or the Advantages and Disadvantages of Being Different

It is interesting, and instructive, to sit in a sidewalk café, and watch the people go by. One is struck by the very great differences in the people walking by. In addition to the most obvious differences in age and sex, people really come in all shapes and sizes. Some of the most obvious differences include weight, height, skin and hair colour, eye shape, bone density, facial bone structure to name just a few. While we do not completely understand the fine detail, we do know that our genes are the main determinants for such differences as hair colour and texture, the colour of our skin and eyes, and facial structure. For other obvious differences such as height, our genes are also believed to be the major determinants although it is known that nutrition also plays a role. Apart from these obvious differences between us, there are many others that are not so obvious but this area of medical research has gained increased attention over the last few years, particularly now that the analysis of the human genome (the complete chemical structure of our genes) has been completed. In particular, scientists are beginning to look at the reasons for the very great variability that exists between individuals. Of particular interest is why some people have an apparently higher risk of developing certain diseases. It does not require qualifications in medicine to realise there is some relationship between our genes and diseases such as cancer, heart disease and diabetes, because many of us know families in which more than one family member has the same disease ie diseases run in families. Doctors have always been aware of this because one of the first questions that they ask a patient presenting with a particular disease is – did your mother/father/brother/sister develop the same disease? What they are really saying is – do you carry the genes that will predispose you to a particular disease?

Most of us would certainly like to know whether, given that we may have a family history of disease, does this mean that our fate is inevitable or can we do anything to prevent or delay the development of disease? Well, there is some bad news and some good news. The bad news is we can do nothing about your genes – we got them from our parents who got them from their parents and so on. However, the good news is that, for the major degenerative diseases that afflict humans, while genes are important, the disease process often requires an environmental – or lifestyle

- trigger(s). In other words, without the environmental trigger, the risk of developing disease is greatly reduced.

Now before we can look more closely at the relationship between our genes, environmental triggers, and disease we need to know something about basic genetics. All of us have over 20, 000 genes that provide a chemical blueprint for the synthesis of over 20, 000 individual proteins. However, we now know that there are many more proteins than genes because the proteins formed from genes can in turn be subject certain changes such the addition of carbohydrate or fatty tags (the products are referred to as glycoproteins and proteolipids respectively) or the nascent proteins may be cut into smaller fragments each with a unique function. For example they can act as hormones, take part in the many steps involved in the absorption and breakdown of our food, the excretion of waste products, the generation of energy and nerve transmission to name just a few. The genes are located on chromosomes, special structures within the cell. Our cells contain 23 pairs of chromosomes. Twenty two of these are termed "autosomes".. One autosome of each pair is derived from our mothers and one from our fathers. In addition, and making up a total of 23 pairs, or 46 chromosomes, females have a pair of special chromosomes, the so-called "X" chromosomes, again one from the mother and one from the father, while males have an X chromosome as well as a Y chromosome, the latter unique to males. In most cases, we inherit two copies of each of the 20, 000 plus genes located on the autosomes, X chromosomes, and Y chromosomes, one from the mother and the other from the father. Females therefore have 22 pairs of autosomes and one pair of X chromosomes, while males also have 22 pairs of autosomes but have to make do with only one X chromosome but also have a Y chromosome. Each of the chromosomes carries a unique assortment of specific genes.

As all of us have the same 20,000 plus genes, apart from the genes that are on the male Y chromosome, which are only found in males, the obvious question is – if we all have the same genes, what is it that makes us all different? The answer is that the chemical structure of the DNA that makes up many of the genes can vary from person to person [1]. In genetic diseases such as hemophilia, the DNA structure in one of the many genes leads to the formation of a protein that is defective and is unable to carry out its normal function at all. This change in DNA structure is termed a "mutation". Unless the person has also inherited a normal gene from one of the parents then the mutation almost invariably leads to disease. In the case of hemophilia, which is caused by an abnormal gene on the X chromosome, there is only a single X chromosome in males and therefore there is no normal gene to counter the abnormal gene so males with the abnormal gene develop hemophilia. Females who have the abnormal hemophilia gene (they are termed "carriers") do not normally develop hemophilia because they also have a normal gene but they will pass on the gene to, on average, one out of every two sons. If an abnormal gene is on one of the other chromosomes, for example the genes that cause cystic fibrosis or phenylketonuria (PKU), the symptoms associated with these conditions will only develop if the person has inherited two abnormal genes, one from the mother and one from the father. These genetic diseases are termed "autosomal recessive' which really means the abnormality is not on one of the X chromosomes but on one of the autosomes whereas conditions like hemophilia are termed 'X linked" genetic diseases.

Of course if that was all, it would be relatively simple. However, there are changes in certain genes, and the inherited brain disease Huntingdon's disease is a good example of this, where the abnormality on a single gene can lead to disease even if the person has also inherited a normal gene from one parent (remember we normally inherit two copies of each gene). In this case, the abnormal Huntingon's gene obtained from one of the parents is able to override or suppress the function of the normal gene inherited from the other parent. People with these sorts of genetic diseases are said to have an autosomal dominant gene ie the abnormal gene is on one of the non-sex chromosomes and it "dominates" the other normal gene. For reasons not understood, in theory at least, disease can develop at any time although in the case of Huntingdon's disease expression of the abnormal gene does not normally occur till at least the third or fourth decade of life. There many of these sorts of diseases, and many others are being discovered all the time. What is really not understood is why the dominant gene causes disease. It has been suggested that abnormal protein formed by the Huntingdon's gene gradually destroys the function of normal brain proteins but why it takes such a long time in many people and less time in others in unclear.

Genetic Polymorphisms and Disease

There is yet another cause of genetic variability which can impact on a person's risk of developing disease and that relates to polymorphisms of genes [3]. We have discussed mutations of genes that can give rise to an abnormal protein product of the gene which is unable to carry out its normal function. We can imagine a situation where a mutation leads to a protein product that is abnormal but in which there is a small amount of activity although insufficient to prevent disease. In fact these types of mutations do exist and may help explain the differences in severity of a particular genetic disease. However, what if the change in the gene was very minor and, to all intents and purposes, the protein product is almost completely normal? These types of changes in DNA, termed "polymorphisms", do exist in many of our genes. However, in much the same way as all houses in a street may have the same function and look much the same yet, in the event of climatic stress such as heavy rain, a gale or an earthquake, some remain undamaged while others suffer serious damage, so too do our genes and their products respond differently to a variety of different environmental stresses such as infections, or chemical exposure, radiation etc. In other words, providing there is little stress, the genes and their products can cope easily but, in certain circumstances, if a particular polymorphism leads to a gene product that is a little less efficient, and the stress is too great, the function of a particular organ may become impaired.

Polymorphisms and Drugs

A good illustration of how polymorphisms can affect the function of our bodies is in the use of drugs to treat disease. Oral drug administration results in its absorption from the gut, its appearance in the blood, gradual removal from the blood, uptake by different organs and in particular by the liver which will either degrade the drug or convert it into a form that is more readily excreted in the urine. All of these processes require gene products and it is easy to see how slight differences in the gene products can have profound effects on rapidity with which the drug is taken up and eventually finds its way, in some form or other, in the urine. Drug administration induces an increase in certain gene products in the liver, termed "cytochrome P450", which begin to act on the drug, changing it chemically to reduce the risk of damage to the different organs and, at the same time, converting it into a form that is more suitable for excretion in the urine. In a way, the induction of cytochrome P450 is the body's way of getting rid of a potentially harmful environmental chemical. Over a period of time, through the action of cytochrome P450, the levels of the drug in the blood begin to diminish and its breakdown products appear in the urine. It is more complicated than this because cytochrome P450 is not a single gene product but in reality there are literally dozens of genes and hence, dozens of gene products. It has been found that the different cytochrome P450 gene products are specific for different drugs and that polymorphisms for many of these different cytochrome P450 gene products exist.

It is now believed that the very great differences observed in the ability of the individual cytochrome P450 gene products in different people to chemically change certain drugs are due to polymorphisms in the various cytochrome P450 genes. It is thought that the up to ten fold difference in the turnover (which measures the decline in blood levels of a particular drug) of many drugs observed in patients or health volunteers is due to polymorphisms affecting the activity of the various cytochrome P450 gene products [2]. These differences are now believed to explain some of side effects of drugs. For example, the drug metoclopramide, also called Maxolon, used to treat nausea, has been reported to cause dystonia (abnormal muscular movements) in some patients due to polymorphisms in one of the cytochrome P450 genes, while polymorphisms in cytochrome P450 have also been blamed for acute kidney damage that develops in some people after treatment with drugs [3, 4]. The influence of polymorphisms on potential damage to tissues caused by drug administration is not confined to drugs that are metabolised by cytochrome P450. A good example of this is the drugs that are used to lower blood cholesterol, the statins. A particular polymorphism of a gene product which plays a role in the clearance of these drugs from the blood (termed "statin transporters") has been found to be associated with an increased risk of one of the main side effects of statins, muscle weakness [5]. It would be surprising indeed if side effects of many other drugs are not shown to be associated with genetic polymorphisms.

But how does this relate to our relative risks in developing degenerative diseases? As mentioned above, there are a number of environmental factors that appear to play a role in the development of disease. These can include certain environmental chemicals, such as pesticides, industrial chemicals, and heavy metals eg mercury and lead. These are taken up into our bodies, and then eventually excreted in the urine or stored somewhere in the body. It is easy to see how the quicker we can get rid of these toxic substances, preferably by excretion in the urine, the less likely they are to cause damage.

Polymorphisms and Risk of Cancer

There are literally hundreds of papers published every year on the question of whether polymorphisms of certain genes in some way increase the risk of cancer. Most of these studies involve comparing the genes of healthy people with those of people with different cancers to see if the cancer sufferers are more likely to have polymorphisms of certain genes which may contribute to the development of their cancer. It is a little like looking for a needle in a haystack because, as discussed earlier, we have more than 20, 000 genes and knowing which of these genes to examine for polymorphisms in the many different cancers is not that easy. However, we do know that there are abnormalities in certain genes in some forms of cancer, most notably breast and ovarian cancers, and colon cancer. The best known examples are the genes referred to as BRCA-1 and BRCA-2.in breast cancer. Certain mutations of these genes, which are inherited in an autosomal dominant fashion, greatly increase the risk of hereditary forms of breast and ovarian cancers. Similarly, around 5–10% of colon cancers are caused by abnormalities in one or other of two gene products, APC or MMR. The latter is involved in DNA repair while the function of APC is not entirely clear. Genetic abnormalities in the most common forms of colon cancer, which have been reported for around 5–10% of all colon cancers, can affect the function of their protein products which are involved in repairing damage to DNA. The abnormalities in BRCA-1, BRCA-2, APC or MMR are sufficient to greatly increase the risk of cancer, in some cases the lifetime risk exceeding 80%. However, these genetic forms of cancer comprise less than 10–20% of all cancers. Most of the other cancers are thought to be caused by so-called "environmental factors", which can mean factors as diverse as viral or bacterial infections, chemical or radiation exposure, smoking, alcohol, or diet. Cancer does not always develop even in those individuals who have, for example BC-1 and BC-2 genes and therefore a higher risk of cancer, so environmental factors are probably involved as well.

Perhaps the most persuasive studies investigating the respective contributions of genetic and non-genetic factors are those carried out on twins and particularly the so-called "monozygotic" twins or identical twins. Monozygotic twins develop from a single fertilised egg (oocyte) which divides into two, each containing identical genes. Each then begins to grow and develop culminating in the birth of two children with the same genetic information. In contrast, the genetic differences between "dizygotic" twins are not all that different from what is found in ordinary brothers and sisters because dizygotic twins develop from two separately fertilised oocytes. By comparing the incidence of individual cancers in monozygotic twins from that in found in dizygotic twins it is possible to determine the relative importance of genetic factors in the development of the disease. A study carried out a number of years ago and published in the prestigious medical journal, Lancet, into the prevalence of death from lung cancer of more than 15,000 male twins, both monozygotic and dizygotic, born between 1917 and 1927, found that the there was little difference in between the numbers of deaths from lung cancer in identical and non-idenitcal twins indicating that genetic factors did not affect the risk of developing lung cancer [6]. It was concluded that environmental factors, in this case the carcinogenic chemicals in cigarette smoke, were the main risk factor.

While many cancers are environmentally induced, and this is clear from studies with identical twins as well as the studies showing great differences in prevalence of individual cancers in different countries, it does not mean that our genes and their products do not contribute. The current view is that it is the interaction of our genes with environmental factors that determines whether we develop cancer or not, and in particular how certain critical genes, many of which contain polymorphisms that can affect the activity of their products, respond to these factors. It is important to emphasise that cancer is not a single disease but is diverse and therefore it could be expected that the respective genetic and non-genetic contribution would vary according to the type of cancer and the organ involved. In fact, this appears to be the case [7, 8]. Indications are that, for example, some of the blood cancers (eg leukemia) and melanoma (a type of skin cancer) have a significant genetic components although even here there are clearly environmental factors involved.

Now what are the environmental factors which can increase the risk of cancer? We know many of them so, with this knowledge, it is theoretically possible to reduce our risk of cancer. They include ultraviolet radiation, perhaps other forms of radiation such as radioactivity, smoking, excess alcohol consumption, asbestos, certain occupational chemicals such as benzene, pesticides, and formaldehyde, and certain dietary choices in excess such as processed meats (the latter containing nitrates and nitrites), smoked foods, and fats, as well as inadequate amounts of foods, such as fresh fruits and vegetables that seem to reduce the risk.

Polymorphisms and Heart Disease

There seems little doubt that environmental factors play a major role in the development of heart disease, and related conditions such as stroke. There is lots of evidence that this is the case. Epidemiological studies (studies involved in comparing the incidence of disease in different population groups) over the last decade or more have repeatedly shown that non-genetic factors are important. Perhaps the most significant of these non genetic factors is diet. Certain fats, eg saturated and trans fats, and perhaps even cholesterol itself have been shown to increase the risk of developing the disease. The evidence for the involvement of non genetic factors is suggested by a number of key observations both on the varying risk of residents of different countries, as well as within different community groups in individual countries. For example, it has been known for many years that people from Mediterannean countries such as Greece have a lower incidence of heart disease but this risk increases when these people migrate to countries (like the US and Australia) where the risk is greater. Further evidence for the involvement of non genetic factors is indicated by studies showing that risk varies within individual countries for certain religious groups, like, for example Seventh Day Adventists (lower risk), vegetarians (lower risk), indigenous minorities (eg Australian aborigines, higher risk) and individuals from different socioeconomic backgrounds (where individuals are put into groups according to social, economic and educational factors). These studies all demonstrate big differences in incidence. Since it would appear unlikely that genetic differences can explain the differences in risk observed between the various groups, it demonstrates that non genetic factors are partcularly important. These non genetic factors include diet, smoking, pollution, lifesyle (lack of exercise), and infections.

Of course this does not mean that our genes do not play a role. It is well known that the incidence of heart disease is increased greatly in people with hypercholesterolemia (which means literally excess blood cholesterol) which is due to a genetic abnormality in a protein which binds cholesterol. Individuals with homocysteinuria (an amino acid) a genetic disease resulting in the accumulation of homocysteine in the blood, have a greatly elevated risk of heart attacks and strokes. In addition to these genetic diseases, there are increasing reports of polymorphisms in different genes. For example, there are indications that there are polymorphisms in a variety of different genes that may increase the risk of heart failure and strokes [9, 10]. There are even suggestions that polymorphisms of certain genes involved in the turnover of statins, one of the major drugs used to bring down blood cholesterol, a risk factor in heart disease, may be associated with some of the severe side effects of the drug observed in some people [11].

What Does All This Mean?

The bad news is that the risk of developing a degenerative disease increases as we get older. However, the good news is that, except for some who have inherited certain abnormal genes that ensure that disease is inevitable, for most of us there is not necessarily an inevitability that we will succumb to diseases like cancer and heart disease in our middle years or in early old age. It has become increasingly apparent that disease mostly develops as an abnormal response of our genes to exposure to certain lifestyle or environmental triggers, many of which are known. Studies with identical twins have demonstrated that while our genes are important, other factors also contribute to our risk and are just as important. It is true that many of us may have inherited certain polymorphisms of one or more of our genes that can predispose us to cancer, heart disease or one or more of the other degenerative diseases, but whether we get sick really depends on two other elements that we can control – environmental factors and lifestyle. It is likely that, in the future, we will have a better understanding of the relationship between our genetic polymorphisms and the disease process and this will give us greater control over our health and wellbeing. In particular, with a knowledge of how polymorphisms in individual genes increase the risk of disease, we will be in a better position to take the necessary action to limit their effects. For example, if we know which environmental chemicals increase our individual risk of disease, we can determine the source of the chemical and make sure that our exposure is reduced. Alternatively, if a lifestyle factor is important, then once we have learnt what that factor is, we can modify our lifestyle.

References

1. Pomomarenko EA et al (2016) The size of the human proteome. The width and depth. Int J Anal Chem 2016
2. Kirchheimer J, Seeringer A (2007) Clinical implications of pharmacogenetics of cytochrome P450 drug metabolising enzymes. Biochim Biophys Acta 1770:489–494
3. van der Padt A et al (2006) Acute dystonic reaction to metoclopramide in patients carrying homozygous cytochrome P450 2D6 genetic polymorphisms. Neth J Med 64:160–116
4. Leung N et al (2009) Acute kidney injury in patients with inactive cytochrome P450 polymorphisms. Ren Fail 31:749–752
5. Voora D et al (2008) The SLCO1BI'5 genetic variant is associated with statin-induced side effects. J Am Coll Cardiol 54:1609–1616
6. Braun MM et al (1994) Genetic component of lung cancer cohort study of twins. Lancet 344:440–443
7. Ahlbom A et al (1997) Cancer in twins: genetic and non-genetic familial risk factors. J Natl Cancer Inst 89:287–293
8. Kadan-Lottick NS et al (2006) The risk of cancer in twins: a report from the childhood cancer survivor study. Pediatr Blood Cancer 46:476–481
9. Guo M et al (2016) Genetic polymorphisms associated with heart failure: a literature review. J Int Med Res 44:15–29
10. Kumar P et al (2016) Role of interleukin-10 ($-1082A/G$) gene polymorphism with the risk of ischaemic stroke: a meta-analyis. Neurol Res 38:823–830
11. Arrigoni E et al (2017) Pharmacogenetic foundations of therapeutic efficacy and adverse events of statins. Int J Mol Sci 18

Chapter 16
Environmental Chemicals and Our Genes

Another part of our bodies that is particularly vulnerable is our genes. As discussed in an earleir chapter, it has been known for quite a while that the "blueprint" or plan for making a human being exists in the part of the cell termed the nucleus, and specifically in tiny structures termed "chromosomes" which in turn are made up of DNA. During the natural turnover or growth of body tissues, each cell in the tissue divides into two, then each of the two divide, and those four also divide and so on. DNA is integral to this process because cell division is preceded by the generation of new DNA using the existing DNA as a template, the DNA in each gene producing an exact copy of itself. During this division of DNA, UV, certain chemicals, or viruses can introduce subtle changes in the chemical structure of DNA. If left unrepaired, these changes can have devastating effects, including cell death and even cancer. Each of us have inherited complicated DNA repair mechanisms in each cell which involve proof reading of the DNA, snipping out bits that should not be there, and replacing them with the correct sequence. Despite this, over time, these repair mechanisms, which themselves are dependent on correct DNA sequences, can also develop abnormalities.

The chemical structure of the DNA is very important because if there is any subtle change, and this can occur naturally within a particular gene, it can result in the generation of an abnormal protein. If the protein is critical to the normal functioning of the cells in tissues such as brain and kidney, then disease can result. Because changes in DNA structure are not that uncommon, there are complicated mechanisms in the body for repairing abnormalities that may develop. Mostly, these processes work exceptionally well but, over time these processes themselves, which are also controlled by genes and therefore dependent on the maintenance of the DNA structure, also develop abnormalities. The importance of these repair mechanisms in the maintenance of health is most apparent in people who have inherited defects in one of the number of genes in involved in this process. These include conditions such as xeroderma pigmentosum, a genetic condition characterised by a hypersensitivity to sunlight which leads to skin cancers and premature aging, and Cockayne syndrome, where a hypersensitivity to sunlight or

UV light and chemical agents results in premature aging and abnormalities in many organs including the brain and eyes.

There is another group of genes that are particularly important in the development of cancer, the tumour suppressors. We know they are important in cancer because if there is an inherited abnormality in one of these genes, cancer can result. Examples include the Li-Freumeni syndrome, a rare condition that develops early in life. Cancers in many different organs are a feature of this condition which is characterised by an abnormality in a tumour suppressor gene, p53. This gene is involved in preventing uncontrolled growth of cells. It is clear that, if left unrepaired, any damage to the p53 gene has the potential to greatly increase the risk of cancer. Many cancers, including lung, breast and colon cancers, show abnormalities in the p53 gene.

How Are Our Genes Damaged?

Whilst DNA is a very complicated substance, it is, in essence, a chemical and, like all chemicals, has the potential to interact with other chemicals to produce something different and this in turn may affect its function. Damage to DNA can be caused by ultraviolet light, radiation, or chemicals. There is a great variety of chemicals that are able to damage DNA. Many of these are said to be "carcinogenic" ie they are able to induce cancer in animals and humans. They do this by reacting with the base components of the DNA. It is a bit of a hit or miss process because, theoretically at least, damage can occur anywhere in our genome (although it is likely that there are "hot spots"). So, if the damage is not repaired, and there may be reasons for this, and depending on which gene is affected and how important it is to the cell, the interaction can cause death of the cell or, in some cases, lead to uncontrolled growth and cancer.

Do Environmental Chemicals Damage DNA?

As DNA is found in almost all cells (red cells, and "cornified" tissue such as nails and skin where the cell nucleus containing the DNA is replaced with the protein keratin, are exceptions), it is predicable that there is a potential for damage in almost all tissues of the body. However, we need to distinguish the so-called somatic (cells not involved in reproduction) from the germline cells (sperm and egg). If damage to DNA occurs in the somatic cells, for example in liver, kidney or brain, then this can lead to loss of function of the cell. On the other hand, if the damage occurs to germline cells, there is an additional possibility and that is, when the egg and sperm unite at the time of conception, the change in DNA may be inherited by the developing foetus. This change in DNA has the potential to cause a miscarriage or increase the risk of birth defects.

There is a great deal of evidence that some environmental chemicals can cause changes in DNA structure. For example, the polycyclic hydrocarbons (PAH), which are pollutants derived from many different sources and in particular from emissions derived from fossil fuels, have been shown to react with DNA in the skin of animals to form modified structures that are believed to explain the development of tumours [1]. PCBs, another group of widely distributed pollutants, have been reported to induce changes in DNA [2]. One of the most active of the polycyclic aromatic hydrocarbons is benzopyrene and its capacity to induce changes in mouse DNA is increased in the presence of another pollutant, arsenic [3]. One of the flame retardants, PBDE 47, has also been shown to induce DNA damage in human cells in vitro (ie in a test tube), an effect that has been shown to increase greatly by the addition of a PCB [4]. This is clearly not a complete list but it illustrates the fact that environmental chemicals do have the capacity to change the structure of DNA, which in turn can affect either somatic or germline (ie egg or sperm) cells.

Is There Evidence That Environmental Chemicals Affect DNA Structure in Humans?

There is increasing evidence that exposure to certain environmental chemicals can lead to changes in DNA structure in humans and this can be readily detected in blood samples. One of the most used method to demonstrate this is the "micronucleus test" [17]. This relies on the principle that, if there is an alteration in DNA structure, it could be anticipated that the change could result in abnormalities in chromosomes, the small bodies within the cell that contain the DNA. When a cell divides, as it does either to replace dead cells or due to growth, the chromosomes also divide so that each new cell contains 23 chromosomes. Some chemicals can actually cut the DNA strand at different points along the DNA chain and this can lead to a break in the chromosome itself which splits into fragments. These fragmented chromosomes can be detected in the red or white blood cells of a blood sample and these cells are then said to contain "micronuclei". The presence of greater than normal amounts of cells with micronuclei are considered to be an indicator of chromosomal and DNA damage [19]. Other tests involve showing the presence of abnormal nucleotides, the building blocks of DNA.

Using these techniques, there have been a number of reports showing that environmental pollutants can induce changes in DNA. Thus children exposed to air pollutants have been observed to have increased numbers of micronuclei in their blood and tissues [5, 16]. Other researchers have found that workers who are exposed to bitumen fumes containing PAH, and other workers exposed to pesticides and formaldehyde respectively, also showed evidence of abnormal chromosomes and DNA damage [6–8]. Finally, urines taken from electronics workers exposed to high levels of PBDEs, PCBs, dioxins, and PAH showed greatly increased levels of 8-HDG, a breakdown product of DNA, a finding consistent with significant damage to DNA in at least some organs of the body [9, 18]. It is important to note, however, that we have mechanisms for repairing damage to DNA. However, there is evidence that even they may be affected by exposure to pollutants [16].

Epigenetics – Non Inherited Changes in DNA

It has been known for a long time that the protein compositions of the many different tissues in our bodies varies quite dramatically even though the same genes, which provide the blueprint for the manufacture of these proteins, are present in all tissues. The explanation for this apparent paradox is that mechanisms for turning on or off the production of individual proteins from their corresponding genes exist in the different parts of our bodies. This explains why body organs so different as, for example, liver, brain and kidney, have completely different structures and functions. DNA analysis demonstrates clearly that each tissue contains the same DNA but that there are minor chemical changes within individual genes, or the structures in which the genes are embedded, which ensure that the expression of the genes ie the production of the corresponding proteins, is either turned on or off, as and when required. As all animals – and humans – develop from a single cell (ie the fertilised egg cell), it is clear that there are mechanisms for turning genes on or off during the development of the unborn baby. Moreover, as gene expression is not confined to the developmental phase alone but occurs at alll stages of life from birth to death, it seems likely that anything that interferes with the process may be harmful.

In most cases these mechanisms are chemical in nature. The most well known is a process termed "methylation" which involves the addition of what chemists term a "methyl group" to the DNA. There is increasing evidence that changes in methylation of individual genes are associated with a disease state [15]. Not surprisingly considering the complexity of the human genome, there are other processes that do not require the use of methyl tags. One of these involves the change in the structure of certain proteins, termed histones, around which DNA is coiled, while yet another involves substances related chemically to DNA, microRNA. Another form of RNA, messenger RNA, is generated from instructions received from the DNA of individual genes and serves as a template for the manufacture of the protein product of the gene but differs in structure to microRNA. When the "methyl group" is added to the DNA it prevents the expression of the gene and therefore the production of the specific protein which is coded for by that gene. If this chemical tag is removed from the gene, a process controlled by yet another protein called a "demethylase", then expression of the gene can take place. In the case of histones, small modifications in chemical structure, can change the three-dimensional structure of the DNA-histone complex either facilitating or suppressing the expression of a particular gene.

It was formerly thought that mechnisms for turning individual genes on or off were inherited at birth and did not change throughout life. However, research carried out over the last decade or so on animals has revealed that new "on/off" gene switches can be created at any time during life, including the period in the womb, and may actually be passed on to the next generation [10, 11]. It is known that the administration of certain chemicals can trigger changes in the DNA, and in particular in the numbers of methyl tags and this may result in disease. For exeample, some of these chemicals include the plastics components BPA and phthalate mentioned earlier in this book [12]. Diethylstilbestrol (DES) is another chemical which has been reported to change some of the epigenetic "tags".This drug was used from the forties into the seventies to reduce the risk of complications in pregnancy. Ir was later shown to increase the risk of rare cancers in daughters of women exposed, and cryptorchidism (undescended testis) and inflammation of the testis in their sons. There is a view that that the effects of DES may be transgenerational ie may even affect the grandchildren of the women treated with the drug [13]. Also, whereas it was thought formerly that epigenetic changes in the developing foetus occurred mainly through maternal exposure to chemicals, there are suggestions that exposure of fathers to chemicals such alcohol and tobacco smoke may have similar effects [14].

What Does All This Mean?

The maintenance of the chemical structure of DNA in our genes is absolutely essential for our survival as a species and this is why we have all inherited complex mechanisms to repair any aberrant DNA structures that may develop during our lives and be transmitted via our genes. At the same time, the maintenance of DNA structure is important for our health and well being because damage can lead to death of the cell, disease, and birth defects. We know that there are a number of ways that DNA in our cells can be damaged but for us the potentially greatest source of concern is the burgeoning number of chemicals developed by humankind and released into the environment. There is no doubt that some of these chemicals can change the structure of DNA. They can do this in different ways but the end result in essentially the same. We do have mechanisms for repairing some of these changes but even these mechanisms, reliant as they are genes and their associated DNA, are also subject to chemical attack. Moreover, even if our repair mechanisms are working efficiently, not all changes in DNA can be corrected. Some of the DNA-damaging chemicals have been identified. However, as more and more chemicals, many of which have not been properly evaluated, are released into the environment, it is likely that many more will be identified in the future. In addition to the effects of pollutants on the chemical structure of our genes, there is increasing evidence that pollutants may also affect the expression of individual genes by changing the composition of the epigenetic tags which function as on/off switches and thereby regulate the expression of genes ie the production of proteins. It is also worth stressing that we really have a limited knowledge of all of the chemicals released and this is because, in some industries, the paper industry is but one example, there are potentially scores of unknown chemical byproducts of the various industrial processes and it is unlikely that at least some of these do not cause damage to DNA.

References

1. Chakravarti D et al (2008) The role of polycyclic aromatic hydrocarbons-DNA adducts in inducing mutations in mouse skin. Mutat Res 649:161–178
2. Mutlu E et al (2016) Polychlorinated biphenyls induce oxidative DNA adducts in female Sprague-Dawley rats. Chem Res Toxicol 29:1335–1344
3. Fischer JM et al (2005) Co-mutagenic activity of arsenic and benzo[a] pyrene in mouse skin. Mutat Res 588:35–46
4. Gao P et al (2009) Influence of PCB153 on oxidative damage and DNA repair-related gene expression induced by PBDE-47 in human neuroblastoma cells in vitro. Toxicol Sci 107:165–179
5. Rossnerova A et al (2009) The impact of air pollution on the levels of micronuclei measured by automated image analysis. Mutat Res 669:42–47
6. Karaman, Pirim (2009) Exposure to bitumen fumes and genotoxic effects on Turkish asphalt workers. Clin Toxicol (Phila) 47:321–326
7. Garai-Vrhovac V, Zeliezic D (2002) Assessment of genome damage in a population of Croatian workers employed in pesticide production by chromosomal aberration analysis, micronucleus assay and comet assay. J Appl Toxicol 22:249–255
8. Costa S et al (2008) Genotoxic damage in pathology laboratory workers exposed to formaldehyde. Toxicology 252:40–48
9. Wen S et al (2008) Elevated levels of urinary 8-hydroxy-2-deoxyguanosine in male electrical and electronic equipment dismantling workers exposed to high concentrations of polychlorinated dibenzo-p-dioxins and dibenzofurans, polybrominated diphenyl ethers, and polychlorinated biphenyls. Environ Sci Technol 42:4202–4207
10. Ho SM et al (2016) Environmental factors, epigenetics, and development origin of reproductive disorders. Reprod Toxicol
11. McLachlan JA (2016) Environmental signalling: from environmental estrogens to endocrine-disrupting chemicals and beyond. Andrology 4:684–694
12. Xin F et al (2015) Multigenerational and transgenerational effects of endocrine disrupting chemicals. A role for epigenetic regulation? Semin Cell Dev Biol 43:66–75
13. Hilakivi-Clarke L (2014) Maternal exposure to diethylstilbestrol during pregnancy and increased cancer risk in daughters. Breast Cancer Res 16:208
14. Day J et al (2016) Influence of paternal preconception exposures on their offspring: through epigenics to phenotype. Am J Stem Cells 5:11–18
15. Radhakrishna U et al (2016) Genome-wide DNA methylation analysis and epigenic variations associated with congenital aortic valve stenosis (AVS). PloS One 11
16. He X et al (2015) Significant accumulation of persistent organic pollutants and dysregulation in multiple DNA damage repair pathways in the electronic-waste exposed population. Environ Res 137:458–466
17. Demircigil GC et al (2014) Cytogenetic monitoring of primary school children exposed to air pollutants: micronuclei analysis of buccal epithelial cells. Environ Sci Pollut Res Int 21:1197–1207
18. Lu SY et al (2016) Associations between polycyclic aromatic hydrocarbon (PAH) exposure and oxidative stress in people living near e-waste recyclic facilities in China. Environ Res 94:161–169
19. Santibanez-Andradea M (2017) Air pollution and genomic instability. The role of particulate matter in lung carcinogenesis. Environ Pollut 229:412–422

Chapter 17
Environmental Chemicals and Mitochondria

The cells of our bodies are a little like miniature cities or towns, with city walls (the membrane or "skin" covering the cell), administrative centres (the cell nucleus), recycling plants (parts of the cell termed "lysosomes"), and factories (the "ribosomes" and "endoplasmic reticulum" are involved in making proteins and other chemicals present in the cell) to name just a few. Every city or town has a power source and, in the case of the cell, these are the mitochondria. Mitochondria are small structures found in most cells in our bodies that are involved in the generation of energy. Mitochondria are able to generate energy from the breakdown of the two main fuels in our food – fat and carbohydrate – and to store it as chemical energy in the form of a complex chemical substance, adenosine triphosphate or ATP. ATP is not unique to animals but is an important energy store in plants, bacteria, and fungi. In essence, ATP production is linked to the removal of electrons, those tiny negatively charged parts of all atoms, from our food, and transferring those electrons along a chain of certain proteins and fats (referred to as the "electron transport chain" or ETC). At various points in this process, the movement of the electrons results in the generation of ATP. The electrons which move along the chain eventually react with oxygen to form water. The vulnerability of this process is due to the fact that minor leakage of electrons can occur and this can result in the formation of what are termed "reactive oxygen species" or ROS. ROS are what are commonly referred to as "free radicals", formed when the electrons react with other components of the mitochondria and the cell, particularly the polyunsaturated fats. Because the ROS are chemically very reactive, there are mechanisms within the cell which destroy them before they can cause too much damage. There are a variety of different substances in our bodies that can eliminate the ROS before they damage delicate cellular structures and these include antioxidants such as vitamin E and C. Despite this, some ROS escape and, over a period of time, the mitochondria are damaged and this in turn can have an effect on the function of the cells in which the mitochondria are located. After all, to continue the town/city analogy, if the power supply is reduced, it will have an impact on city maintenance.

A. Poulos, *The Secret Life of Chemicals*,
https://doi.org/10.1007/978-3-030-80338-4_17

While most of the genetic material in the form of DNA is located in the nucleus of the cell, mitochondria are unique in that they are the only other part of the cell that contain DNA. Unlike the DNA which is in the nucleus and which provides the chemical blueprint required to make proteins in any part of the cell, mitochondrial DNA provides the blueprint for making only proteins found in mitochondria.

Because of the key role of mitochondria in energy production, it follows that anything that interferes with this process, can result in the death of the cell, the corresponding shut-down of a whole organ such as the brain, kidney, or liver, and death of the organism itself, whether it is human, animal or bacterial. For example, inherited abnormalities in mitochondrial genes can lead to loss of function of mitochondria and this in turn can result in diseases affecting different parts of the body, particularly muscle and the brain [5, 6, 10, 11].

Chemicals are known to interfere with the function of mitochondria. Perhaps the best known of the mitochondrial poisons is cyanide which works by blocking the movement of electrons along the ETC but there are many other mitochondrial poisons. Two of the best known mitochondrial poisons are environmental chemicals, eg pesticides. One of these is rotenone, present in the roots and stems of various plants and best known by organic gardeners as a component of derris dust with insecticidal properties. The other is paraquat, one of the most used herbicides [1, 2, 8, 9]. Both of these substances produce their effect by stimulating the production of ROS. The toxic heavy metals arsenic, cadmium and mercury have also been reported to increase ROS production via their effects on mitochondria and it has been suggested that this could explain their ability to produce abnormalities in brain function [2, 7]. Rotenone and paraquat in particular have been shown to produce Parkinson's disease-like symptoms in animals and also induce the gastrointestinal abnormalities (delay in gastric emptying) that are also a feature of this disease [3, 4].

In addition to the mitochondrial poisons mentioned above there are others and they include the quite well known antiobiotics such as oligomycin and antimycin, dequalinium chloride which has been used as an antiseptic in wound dressings and mouth infections, 2, 4-dinitrophenol used many years ago as a diet pill, and MPTP. The latter was discovered after some users of the synthetic opioid (a chemical derivative of opium) meperidine developed Parkinson's disease symptoms. This was later shown to be due to the presence of a contaminant, MPTP. While it is clear that there are already many chemicals with the capacity to induce abnormalities in mitochondria and hence destroy the function of organs such as the brain, there is little doubt that many more probably exist amongst the thousands of industrial chemicals, some of which have been mentioned in earlier chapters of this book, which are released into the environment every day and find their way into our bodies. Most are not tested for their effects on mitochodria. It is assumed that most of the pollutants released into the environment will be broken down by microorganisms in soil or water, or chemically degraded through the action of water, heat, or UV radiation, and hence will not pose a threat to fauna or humans. This assumption may not necessarily apply for all of thousands of pollutants produced from human activities.

What Does All of This Mean?

Most people know little about mitochondria. This lack of knowledge about mitochondria is surprising given the great importance of these tiny structures in almost all of the major organs of the body. We know how important they are because of the existence of certain genetic diseases caused by an abnormality in certain mitochondrial components. Also, certain substances, such as cyanide, can interfere with mitochondrial function and this can rapidly result in death. Despite their importance, they have a certain vulnerability which is a byproduct of their function ie to generate chemical energy from the food we eat. The movement of electrons from the food components to oxygen is the cause of this vulnerability because leakage can occur during this process and, if unchecked, can result in the generation of highly active free radicals that react with almost anything in the cell. These free radicals can damage components of the ETP, as well as mitochondrial DNA. If this damage is not corrected, or only partly corrected, then these abnormalities can lead to the loss of function of mitochondria and this in turn can affect the function of the cell in which the mitochondria are located. There are processes that can correct abnormalities as they develop, for example the deactivation of free radicals by certain proteins, or antioxidants such as vitamin E or C, or certain proteins, can destroy the antioxidants before they cause any damage. In addition, if the free radicals cause changes in mitochondrial DNA, then there are special proteins which we all have that can repair the damage. Despite this, mitochondrial abnormalities do develop over time and it is likely that environmental factors may contribute to this. Certain chemicals are known mitochondrial toxins, for example pesticides like rotenone and paraquat, and it is likely that some of the many thousands of environmental chemicals, most of which have not been tested, may also be mitochondrial toxins and therefore may contribute to degenerative diseases such as cancer, diabetes, and heart disease.

References

1. Xiong N et al (2012) Mitochondrial complex 1 inhibitor rotenone induced toxicity and its potential mechanisms in Parkinson's disease models. Crit Rev Toxicol 42, 613–632. Corrigan FM et al (2000) Organochlorine insecticides in substantia nigra in Parkinson's disease J Toxicol Environ Health A 59, 229–34
2. Monroe RK, Halvorsen SW (2009) Environmental toxicants inhibit neuronal Jak tyrosine kinase by mitochondria disruption. Neurotoxicology 30:589–598
3. Cannon JR et al (2009) A highly reproducible rotenone model of Parkinson's disease. Neurobiol Dis 34:279–290
4. Somavayajulu-Nitu et al (2009) Paraquat induces oxidative stress, neuronal loss in substantia nigra region.and parkinsonism in adult rats: neuroprotection and amelioration of symptoms by water soluble formulation of coenzyme Q10. BMC Neurosci 10:88
5. Schapira AH (2012) Mitochondrial diseases. Lancet 379:1825–34
6. Das NR, Sharma SS (2016) Cognitive impairment associated with Parkinson's disease: role of mitochondria. Curr Neuropharmacol 14:584–592

7. Prakash C et al (2016) Mitochondrial oxidative stress and dysfunction in arsenic neurotoxicity: a review. J Appl Toxicol 36:179–188
8. Blanco-Avala T et al (2014) New insights into antioxidant strategies against paraquat toxicity. Free Radic Res 48:623–640
9. Huang CL et al (2016) Paraquat induces cell death through impairing mitochondrial membrane permeability. Mol Neurobiol 53:2169–2188
10. Grimm A, Eckert A (2017) Brain aging and neurodegeneration: from a mitochondrial point of view. J Neurochem
11. Alston CL et al (2017) The genetics and pathology of mitochondrial disease. J Pathol 241:236–250

Chapter 18
Environmental Chemicals and the Immune System

Another vulnerable part of the body is the immune system particularly in the first year of life. Its activity also declines with age. The human immune system is a highly complex and coordinated structure. Its primary goal is to ensure the survival of an individual in the event of bacterial, viral, fungal or parasitic attack. Apart from its role in combating infections, its various cellular components seek out and destroy cancer cells, and are involved in the repair of damaged tissue or organs. To produce its effects it is reliant on special proteins such as antibodies and other proteins termed cytokines that can increase the activity of certain cellular components of the immune system, such as the natural killer cells, that can kill cancer cells as well as cells infected with viruses. The immune system is divided into two arms – the innate and adaptive systems. The innate system represents a non-specific first line of defence against infection and includes the natural killer cells, as well as other cells such as neutrophils and macrophages, which engulf and destroy infective agents. The cells of the innate immune system have, at their disposal, a veritable armory of potent chemicals that can kill any infecting organisms. The adaptive immune system is comprised of special types such as B-cells and T-cells that also play a role in combating infections and cancer but this arm of the immune system is normally slower to respond. However, where the two arms differ is that, after they have responded to a particular infection, they retain a memory of the infecting organism so that, at some future date, if a person is exposed to the same organism, the response from the T-cells is much more rapid.

The cells of the immune system are mostly generated in the bone marrow from the so-called stem cells (they have the potential to produce any cell type found in blood) and migrate throughout the body where they play different but coordinated roles in immunity. Some organs, notably the spleen, thymus, and lymph nodes, have important accessory functions. For example, immature cells from the bone marrow travel to the thymus where they undergo a maturation process that converts them into different types of T-cells. The spleen is involved in the turnover of old or damage red blood cells from the circulation, but is also the site of the production of antibodies also made by immune cells, while lymph nodes serve as filters of bacteria and other infectious agents in the blood and the lymph. The various immune cells generally have to be "activated" to produce their effects. Thus, chemical signals released from bacteria trigger the movement of cells of the innate immune system towards the bacteria and other signals "prime" these cells ie they greatly increase the response towards the infecting agent. Similarly, cells of the adaptive immune system, for example T-cells, are also primed by chemical signals. Some of these signals are derived from bacteria, others are the proteins referred to above, the cytokines. There are a great number of these cytokines produced by immune cells and they mostly work by locking on to special receptors on the surface of the cell and this in turn starts a cascade of reactions that result in a particular response. Just to give one example, the interaction of certain cytokines with one particular cell type, the neutrophil, ultimately generates large amounts of a substance "superoxide" that is highly reactive and is able to kill any bacteria that are engulfed by the neutrophil.

While the "activation" of immune cells is critical for their function, it is also critical that, once they have carried out their task, they are "deactivated" otherwise the activated cells can play havoc on any tissue that is nearby. So there are complex systems for damping down these activated cells and, again, these are controlled by cytokines. If this system does not work efficiently then damage to certain parts of the body can occur. If the damage occurs in joints, then arthritis will result. The function of whole organs can also be compromised. For example, the juvenile form of diabetes or type 1 diabetes is thought to be due to an immune system attack of the gland that makes insulin, the pancreas. There are literally dozens of diseases that seem to be associated with an overactive immune system leading to a loss of function of a particular organ. These diseases are referred to as inflammatory diseases and they mostly involve a chronic, as opposed to an acute, involvement of the immune system. Chronic inflammation may occur over many years, often with no signs in the early stages, until eventually the function of the organ or joint is affected. Asthma, which affects the lungs, and multiple sclerosis, affecting the white matter in the brain are just two examples. Even coronary heart disease is now believed by some scientists to result from chronic inflammatory process affecting the walls of the coronary arteries.

It is fairly obvious that anything that interferes with the delicate balance of activation and damping down can have potentially devastating effects on our health and wellbeing. There are number of possibilities as to how this may occur. For example, it is known that there are a number of inherited genetic diseases which are caused by abnormalities in the genes that affect one or more of the many protein components of the immune system. The best known example is the "boy in a bubble" disease, or combined immunodeficiency. People with this disease are unable to make antibodies, a key component of the immune response, and this results in ongoing bacterial and viral infections. Another has the rather complicated name -chronic granulomatous disease -caused by an abnormality in a protein present in neutrophils which is involved in the generation of superoxide, essential for the killing of bacteria [1]. People with these and other genetic immundeficiency diseases are prone to developing serious infections. The immune system can also be compromised in non-genetic diseases, the best example being AIDS, where the HIV virus attacks T-cells and this can increase the risk of infections.

Chemical agents can also interfere with this process. There are various well known drugs that can inhibit the immune response, perhaps the best examples are the steroids, cortisone in particular, and cyclosporine. Cortisone can depress both arms of the immune system and is used to treat inflammation in general, while cyclosporine, an antibiotic derived from fungi, has been used extensively to reduce the risk of rejection in organ transplants but there are literally dozens of natural and synthetic chemicals that can act as immunosuppressants [2]. Perhaps one of the best known are the fish oils and the omega-3 fats they contain.

Other chemical substances can act as adjuvants ie they are able to enhance the response of the immune system and are often components of vaccines. Some of these are relatively simple substances, like certain aluminium salts, squalene (a fatty substance found in certain oils such as olive oil and shark liver oil), and isoprinosine. There are also simple peptides (proteins are peptides and are made up of numbers of amino acids joined together) with relatively small numbers of amino acids. One of these is tuftsin (with four amino acids) [3]. One of the best known of the larger peptide stimulators of the immune system is transfer factor, made up of a great number of amino acids and present naturally in a variety of natural sources including egg yolk, white blood cells, and colostrum. Yet another is flagellin, a peptide component of the flagellae (a structure that enables movement) of bacteria [4].

While an increase in immune activity, such as that which occurs during a bacterial or viral infection may be beneficial, in other situations it may be harmful. For example, we now know that there are a great number of environmental and occupational chemicals which are able to induce an increased immune response but this can result in the inflammation found in conditions such as asthma and dermatitis. Perhaps the most devastating consequences of an overactive immune system is the response of some people to substances in certain foods, certain proteins in peanuts are a good example, which trigger anaphylaxis and, in some cases, death. There is thus little doubt that certain naturally occurring, as well as man made chemicals, can have quite profound effects on the immune system [5].

How Do Environmental Chemicals Affect the Cells of the Immune System?

Given that we know that chemicals have been shown to be able to regulate the activity of the immune system, what about environmental chemicals? As there are now thousands of these chemicals, surely it is unlikely that they are all immunologically inert? And if they are not inert, what do they do? There is now a great deal of published information on how some of these can affect the various elements of the immune system. Much of this information relates to how the various chemicals affect human immune cells in vitro ie quite literally in a test tube. These experiments involve obtaining human or animal cells, mostly, although not always, isolated from blood, adding the chemical that is being investigated, and examining the response of the cell.

There is now clear evidence that one of the major classes of environmental chemicals, the organochlorine pesticides, can affect the function of natural killer cells. These cells play an important role is seeking out and destroying both cancer cells and cells infected with viruses. They do this by approaching a cancer cell or virus infected cell and releasing certain substances that penetrate the offending cell, triggering a process that is referred to as "apoptosis". Apoptosis is a type of regulated cell death. In the case of viral infected cells, the big advantage of apoptosis is that it does not lead to the death of the cell followed by the concomitant release of viral particles that can re-infect other cells. Rather, the cell itself kills off the viral particles. Researchers have found that if natural killer cells are exposed to certain organochlorine pesticides their ability to kill cancer cells is greatly diminished [6]. Other studies have indicated that other classes of pesticides, the organophosphorus insecticides and fungicides, also have the capacity to kill natural killer cells [7, 8]. These effects of pesticides are not confined just to natural killer cells because there have also been report of similar effects on T-lymphocytes, the major cell type of the adaptive immune system [9, 10].

These effects are really not the exception because there are many other environmental chemicals that can also affect immune cells. For example, formaldehyde, mentioned in an earlier chapter, and used in a number of industries, has been shown to have very interesting effects on the immune system. In one report, it was observed that human T-cells exposed to formaldehyde displayed a reduced production of certain key proteins involved in regulating the immune response (interleukin-10 and interferon-gamma [11], Diesel exhaust particles, containing a mixture of chemical pollutants, produced a similar effect. Yet another group of researchers found that occupational exposure to formaldehyde decreased the number, and also changed the proportions of the different types of immune cells [12]. In mice, exposure to formaldehyde appears to enhance the response of the immune system to allergens [13]. The latter are substances that trigger responses of the immune system, in some cases the effects are not serious but uncomfortable (eg hay fever) but in others potentially life threatening (eg asthma, peanut allergy).

T-cells in humans and animals are comprised of two or more types. The two main types are termed t-helper 1 and t-helper 2, the former being mainly involved in regulating the activity of the cells of the innate immune system (eg macrophages) while the t-helper 2 cells stimulate the production of antibodies. There is a delicate balance between these two types and anything affecting this can have profound effects on health. For example, there are indications that alterations in the balance of these two subsets of T-cells can influence the outcome of lung infections, and the development of kidney abnormalities in inflammatory diseases such as lupus [14, 15]. It is known that certain environmental chemicals can affect this balance. For example, some PCBs, the environmentally stable chemicals that have a wide distribution and discussed earlier in this book, are able to change the relative proportions of these two types of T-cell in cultured mouse cells [16]. Differences in the two T-cell populations exposed to some PCBs have also been observed in cells isolated from the blood of human volunteers [17].

The function of another important immune cell type, the macrophage (cells of the innate immune system) which works by engulfing sources of infection and then destroying them through the release of digestive substances, is also affected by certain pesticides [18, 19]. The effect on macrophages is not confined to pesticides but has also been observed for other environmental chemicals, for example hydroquinone, a chemical derivative of benzene, and it would indeed be surprising if at least some of the many thousands of environmental chemicals do not interfere with the function of macrophages and other cells of the innate immune system [20].

Is There Any Evidence That Environmental Chemicals Affect the Immune System of Animals?

It is now fairly clear that environmental pollutants are present in wildlife such as birds, reptiles, mammals and fish [21, 22]. Even the tissues of animals from some of the most remote parts of the planet have now been found to contain a variety of pollutants. For example, a study carried out on tissues taken from polar bears in the Arctic have confirmed the presence of PCBs, PBDEs, fluorocarbons and organochlorines [23]. The levels of some of the pollutants found in marine mammals are now believed to be approaching those thought to be toxic to the immune system [30]. There is plenty of evidence that exposure to chemical pollutants can affect wildlife such as birds, reptiles, mammals, and fish [24]. Experiments carried out with mice exposed to individual pesticides have demonstrated a reduction in activity of the immune system [25]. In addition, there are suggestions that exposure of mice to combinations of environmental chemicals such as pesticides may produce a greater reduction in activity than shown by a single chemical [26]. All of these reports confirm that the immune systems of many different animals, including mammals, are affected by environmental chemicals.

Is There Any Evidence That Environmental Chemicals Affect the Immune System of Humans?

It is one thing to show that environmental chemicals can interfere with immune function of animals or even human cells in a test tube, but quite another to demonstrate an affect on people exposed to these same chemicals. There is now increasing evidence, and reported in medical and scientific journals, that this is the case. Just a few examples will suffice.

Mention has already been made in this book of the effects of the ability of some of the chemical substances in dusts produced in mining to induce inflammatory changes in the lung and this in turn can cause disease. Examples of this include coal miners' black lung, asbestosis, silicosis, and chronic beryllium disease have already been mentioned. These diseases are all due to an abnormal immune response. There are indications that exposure to other chemical pollutants may affect the function of the human immune system. For example, an increased risk of asthma has been reported to be associated with exposure to chemicals in the workplace [27, 28]. Chronic inflammation leading to abnormalities in lung function including bronchitis and asthma have also been associated with occupational exposure to pesticides [29, 30]. Exposure to a variety of chemicals can cause dermatitis, or skin inflammation, a condition believed to affect 10–15% of the population in the developed world [31, 32].

There are also reports of a relationship between exposure to dioxins, PCBs, and pesticides and rheumatoid arthritis in women [33, 34]. Rheumatoid arthritis, a disease caused by degeneration of various joints in the body, is thought to result from abnormalities in the immune system and specifically in the attack, by the immune system, of parts of the body, in particular joints.

What Does All This Mean?

The importance of our immune system in the maintenance of health and wellbeing cannot be overstated. It defends us from infections caused by bacteria, viruses, parasites and fungi, seeks out and destroys cancer cells, and helps in the healing processes that are a part of any infection. We take for granted this marvellous system that has gradually developed through evolutionary processes over hundreds of thousands of years. However, like any other part of our body, it can be compromised by exposure to some of the thousands of chemicals to which we are exposed in our day to day lives. This is not surprising because the regulation of the immune system is controlled by chemicals, albeit natural chemicals produced by our own bodies or present in our food. We know some of the chemicals that can affect our immune system from our studies with animals, from the effects produced on human immune cells in a test tube and, in some cases, from their effects on people. What we do not know is how many of the diseases that affect us are related to the exposure of our immune systems to environmental chemicals. Given that environmental chemicals have been clearly shown to affect many different immune cell types, and given also that there are literally thousands of chemicals released into the environment or added to our food, many of which are almost certainly not tested for potential immunotoxic effects, it would seem most unlikely that they are all totally inert and therefore without effect. Of course, government authorities would argue, and quite convincingly, that the amounts that find their way into our bodies are too small to have any effect and, anyhow, exposure has to be balanced against the positive contribution many of the industrial chemicals have on our way of life. In other words, some disease is the price we have to pay for progress. One response to this is to argue that we do not know the full extent of the problem and, if we did, we may argue that it is not acceptable. Moreover, it is easy to be dispassionate if one, or one's family members or friends, are not personally affected. The view that we are only exposed to very small amounts and therefore there is no problem has been explored repeatedly in this book. Unfortunately, small does not necessary mean safe and, after all, all of the testing (assuming that there has been testing) has been with animals, and it is rare for any animals to be tested with the great variety of chemicals to which we are exposed.

References

1. Roos D (2016) Chronic granulomatous disease. Br Med Bull 118:50–63
2. Laev SS, Salakhutdinov (2015) Anti-arthritic agents: progress and potential. Bioorg Med Chem 23:3059–3080
3. Najjar VA Tuftsin, a natural activator of phagocytic cells: an overview. Ann NY Acad Sci 419:1–11
4. Honko AN, Mizel SB (2005) Effects of flagellin on innate and adaptive immunity. Immunol Res 33:83–101
5. Quirce S, Bernstein JA (2011) Old and new cases of occupational asthma. Immunol Allergy Clin Norht Am 31:677–698
6. Reed A et al (2004) Immunomodulation of human natural killer cell cytotoxic function by organochlorine pesticides. Hum Exp Toxicol 23:4630471
7. Li Q et al (2007) Organophosphorus pesticides induce apoptosis in human NK cells. Toxicology 239:89–95
8. Li Q et al (2014) Carbamate pesticide-induced apoptosis and necrosis of natural killer cells. J Biol Regul Homeost Agents 28:23–32
9. Li Q et al (2015) Carbamate pesticide-induced apoptosis in human T lymphocytes. Int J Environ Res Public Health 12:3633–3645
10. Li Q et al (2009) Chlorpyrifos induces apoptosis in human T-cells. Toxicology 255:53–57
11. Sasaki Y et al (2009) Molecular events in human T cells treated with diesel exhaust particles or formaldehyde that underlie their diminished interferon gamma and nterleukin 10 production. Int Arch Allerg Immunol 148:239–250
12. Hosgood HD et al (2013) Occupational exposure to formaldehyde and alterations in lymphocyte subsets. Am J Ind Med 56:252–257
13. Gy Y et al (2008) Long term exposure to gaseous formaldehyde promotes allergen specific IgE mediated immune responses in a murine model. Hum Exp Toxicol 27:37–43
14. Pinto RA et al (2006) T helper 1/T helper 2 cytokine inbalance in respiratory syncytial virus infections is associated with increased endogenouse plasma cortisol. Pediatrics 117:878–886
15. Tucci M et al (2008) Overexpression of interleukin-12 and T-helper 1 predominance in lupus nephritis. Clin Exp Immunol 154:247–254
16. Sandal S et al (2005) Effects of PCBs 52 and 77 on the Th1/Th2 balance in mouse thymocyte cultures. Immunopharmacol Immunotoxicol 27:601–613
17. Gaspar-Ramirez O et al (2012) Effect of polychlorinated biphenyls 118 and 153 on Th1/Th2 cells differentiation. Immunopharmacol Immunotoxicol 34:627–632
18. Helali I et al (2016) Modulation of macrophage functionality induced in vitro by chlorpyrifos and carbendazim. J Immunotoxicol 13:745–750
19. Ustyugova IV et al (2007) 3, 4 –Dichloropropionaniline suppresses normal macrophage function. Toxicol Sci 97:364–374
20. Lee JY et al (2007) Hydroquinone, a metabolite of benzene, reduces macrophage-mediated immune responses. Mol Cells 30(23):198–206

21. Batt AL et al (2017) Statistical survey of persistent organic pollutants: risk estimations to humans and wildlife through consumption of fish from US rivers. Environ Sci Technol 51:3021–3031
22. Hamlin HU, Guilette LJ (2010) Birth defects in wildlife: the role of environmental contaminants as inducers of reproductive and developmental dysfunction. Syst Biol Reprod Med 56:113–121
23. Dietz R et al (2015) Physiologically-based pharmacokinetic modelling of immune, reproductive and carcinogenic effets from contaminant exposure in polar bears (Ursus maritimus) across the Arctic. Environ Res 140:45–55
24. Hamlin HU, Guilette LJ (2010) Birth defects in wildlife: the role of environmental contaminants as inducers of reproductive and developmental dysfunction. Syst Biol Reprod Med 56:113–121
25. Fukuyama T et al (2013) Immunotoxicity in mice induced by short-term exposure to methoxychlor, parathion, or piperonyl butoxide. J Immunotoxicol 10:150–159
26. Nishino R et al (2014) Effects of, short-term oral combined exposure to ennvironmental chemicals in mice. J Immunotoxicol 11:359–366
27. Kogevinas M et al (2007) Exposure of substances in the workplace and new onset asthma: and international prospective population-based study (ECRHS-II). Lancet 370:336–341
28. Baur X (2013) A compendium of causative agents of occupational asthma. J Occup Med Toxicol 8:15
29. Zuskin E et al (2008) Respiratory function in pesticide workers. J Occup Environ Med 50:1299–1305
30. Mamane A et al Occupational exposure to pesticides and respiratory health. Eur Respir Rev 24:306–319
31. Zug KA et al (2009) Patch-test results of the north American contact dermatitis group 2005-2006. Dermatitis 20:149–160
32. Karlberg AT et al (2008) Allergic contact dermatitis- formation, structural requirements, and reactivity of skin sensitisers. Chem Res Toxicol 21:53–69
33. Parks CG et al (2011) Insecticide use and risk of rheumatoid arthritis and systemic lupus erythematosus in the Women's health initiative observational study. Arthritis Care Res (Hoboken) 63:184–194
34. Lee DH et al (2007) Positive associations of serum concentration of polychlorinated biphenyls or organochlorine pesticides with self reported arthritis, especially rheumatoid type, in women. Environ Health Perspect 115:883–888

Chapter 19
Just Because the Amounts Are Small, Does It Mean They Are Safe?

Most people would be surprised to learn how little of some of the key nutrients in our food are required on a daily basis. Around 2-3 micrograms (a microgram is a millionth of a gram) of vitamin B12 is enough to keep us from developing a form of anaemia, 5 micrograms is sufficient to prevent rickets (a bone disease) in children, and 10–20 micrograms of vitamin K is enough to prevent very young children from bleeding. Conversely, people would also be surprised learn how little of certain poisons are required to cause illness. For example, one microgram or less of the botulism or tetanus toxins is sufficient to cause harm or even death. There are, as well, other more common examples of how small amounts of certain chemicals have the potential to cause harm or even death and these include the allergies to certain foods such as peanuts, egg or milk where a few milligrams (thousandths of a gram) are sufficient to cause a response. The actual amount that triggers the response in these cases is much less than this, perhaps in the tens of microgram range, because the active component in the food, mostly a protein, may represent a small proportion of the food components. There is therefore no question that amounts that we consider exceedingly tiny do have the potential to keep us healthy or even kill us. It all depends on the nature of the chemical and the various bodily processes it affects. So, while small can be beautiful, it can also be lethal.

There is also no doubt that some of the chemicals we are exposed to in our every day lives are harmful to us if taken in large enough amounts. Indeed, one could argue that just about everything, even oxygen and ordinary salt are poisonous if the amounts taken are sufficiently large. Certainly, many of the pesticides, metals such as arsenic, lead and mercury, and a great number of the industrial waste products are considered toxic and yet governments believe that, in the amounts were are exposed to, they do not pose a risk to our health. However there are numerous examples of chemicals causing disease when taken in amounts much lower than the poisonous dose.

Perhaps the best example is smoking. While there are dozens of potentially poisonous and cancer causing substances in cigarette smoke, the amounts a smoker is exposed to in each cigarette are much lower than that required to harm. And yet

A. Poulos, *The Secret Life of Chemicals*,
https://doi.org/10.1007/978-3-030-80338-4_19

research has demonstrated fairly clearly that there is a link between smoking and lung cancer in particular. Another example is the toxic metal arsenic. In large doses (from a few hundred milligrams to gram amounts it kills – the actual amount depending on the chemical form of the arsenic - but amounts a hundred fold or more lower than this can also cause skin abnormalities, nerve damage, and anaemia. Similarly, while a few grams of mercury in its different forms can kill outright, much smaller amounts can affect the brain, kidney and other organs. However, in order for a small amount to be harmful, such as in, for example, cigarettes, it has to be taken regularly over long periods of time perhaps even decades.

The main reason for this – and this applies to most environmental chemicals – is the fact that the amounts that eventually find their way into our bodies are so minute that it is difficult to see how they can be harmful at these levels. There is a problem with this argument because there are numerous everday examples where small, non poisonous doses of a chemical do cause disease. It is worth looking again at how toxicity is assessed. The normal way of determining whether a chemical is potentially toxic is to try it out on animals, the most used species being rodents such as rats and mice. For these experiments, animals are fed steadily increasing amounts of a chemical until a toxic dose, referred to as the LD50 or the amount which results in the death of 50% of all animals is determined. This figure establishes the likely toxicity of a particular chemical in humans. The higher the LD50, the less toxic the chemical. It has always been believed that, providing the amounts of chemical we are exposed to are kept perhaps a hundred or a thousand times lower than this figure then there should be no effect on us. On the surface at least, this does not appear to be an unreasonable arguement.. However, there are a number of problems with these sorts of studies. Firstly, rats and mice are not humans and, despite the fact that there are similarities in the way rats and humans handle chemicals such as PCBs, there are differences as well. Secondly, most toxicology studies, particularly for environmental chemicals such as PCBs, do not involve very long term chronic exposure which can more accurately mimic the exposure of humans. The situation is different with pharmaceuticals where toxicological testing is much more rigorous and, moreover, even after approval for a particular drug, once the drug is released into the market there is continuing monitoring. It is really impractical to subject the thousands of industrial chemicals to this degree of testing and scrutiny. Thirdly, the studies carried out with rodents do not take into account the fact that many humans have pre-existing conditions (diabetes, cancer, arthritis, hepatitis, nephritis etc), or smoke, or drink excessively, or take recreational drugs. These factors can affect the capacity of an individual to deal with a chemical, perhaps even at the very low concentrations that may be present in our food, water or air. And finally, it must be emphasised that any toxicological testing that is carried out, even for pharmaceuticals, almost always involves the testing of a single substance. It is rare that mixtures of chemicals are used. However, in the case of environmental chemicals, it is the rule rather than the exception, that we are exposed to complex mixtures. Irrespective of what governments may tell us, there is no way one can know for certain what these complex mixtures may do to our health.

The New Toxicology Paradigms – Hormesis

The present, and past, methods of assessing toxicity are of limited value when dealing with chronic exposure over many years. Testing for toxicity resulting from exposure over many years is possible in animals but the rodent models do not take into account the much longer life span, and therefore the longer exposure, of humans as compared to rodents. There are now real doubts about the relevance of animal testing and what it can tell us about long term exposure of humans to small amounts of toxic substances. Whereas in the past toxicologists (they are scientists who study poisonous substances) believed that there was a concentration threshold below which poisons had no effects on animals or humans, scientists working in an area of toxicology known as "hormesis" have espoused the view that sub-lethal amounts of a poison may also produce some effect, perhaps even an opposite effect. Perhaps because this view smacks somewhat of homeopathy, an alternative way of treating disease and largely dismissed by many scientists, it has been controversial. However, in the last decade there have been numerous reports in journals that point to measurable effects at very low and sublethal amounts of a toxic substance.

A couple of examples will suffice and these concern cadmium and glyphosate. Cadmium is a known toxin and chronic exposure can lead to bone and kidney diseases. Acute exposure to cadmium leads to severe gastrointestinal problems and severe lung inflammation ultimately leading to death. One of the major effects of chronic exposure is damage of the kidneys. Experiments carried out using lung cells taken from human embryos actually showed two effects – one at low concentrations that stimulates the growth of these cells, and an inhibition of cell growth at high concentrations [1] Glyphosate is a well known herbicide. However, there is increasing evidence that in much smaller amounts, such as may occur in spray drift onto non-sprayed fields, it may stimulate growth of some plants [2] These two examples, and there are many others, point to a phenomenon that is truly surprising. In the case of the examples cited, it indicates that very small amounts of something that is toxic in larger amounts may be beneficial. Of greater importance, however, is the conclusion that very small amounts of substances, both toxic and non-toxic, may not necessarily behave in a manner expected. And, further, if toxic substances may have certain beneficial effects at very low levels, can they also have other, as yet unrecognised, harmful effects?. After all, in the examples cited, and in much of the available literature on the topic of hormesis, the focus has been on the systems known to be affected by large doses of a toxin without any consideration of the scores of other systems or pathways that may be vulnerable.

Synergism

Another way that tiny amounts of chemicals can produce an effect is via the process of synergy where the combined activity of two separate substances is much greater than than that predicted by adding the sum of each. This is a well known phenomenon in medicine where mixtures of drugs can apparently produce unexpected, and often harmful, effects that are not predictable from a knowledge of the effects of each individual drug. There are also many examples of synergy using mixtures of pesticides against insects or fungi. Perhaps the most impressive studies are those showing synergy between rotenone, an insecticide found in derris dust, and lipopolysaccharide (LPS), parts of the cell wall of many bacterial strains. LPS occurs naturally in humans because antibodies to LPS are routinely detected in blood and very small amounts, of the order of fractions of a millionth of a gram, can provoke inflammation at quite low concentrations. Rotenone, as discussed earlier in this book, is a mitochondrial poison and works by stimulating the production of free radicals and this can damage the mitochondria, the tiny powerhouses of the cell, ultimately leading to the death of the cell. A group of US researchers, studying the effects of rotenone and LPS on certain brain cells that specialise in the production of dopamine, an important substance involved in the transmission of nerve impulses in the brain, found that while rotenone, by itself, at concentrations much lower than the toxic dose, produces an apparently negligible effect on the brain cells, in combination with LPS (again at levels below that required to produce any effect), can induce a large increase in free radical formation ultimately leading to the death of cells [3]. This is an interesting observation with relevance to Parkinson's disease, a degenerative disease affecting humans, which is caused by a gradual loss of function of brain cells that make dopamine. The researchers speculated that while the causes of Parkinson's disease are not known, it is possible that the disease may result from the effects of interactions among multiple factors, another way of saying that synergism may be involved.

Priming

There is yet another way tiny sub-lethal amounts of a substance can produce marked and unexpected effects and that is via a mechanism that is referred to as "priming". There is some overlap between the mechanisms of hormesis, synergy and priming. However, priming is a biological process that is well known to scientists, particularly those who work in the field of immunology. Some of the special proteins produced by the immune system, the so-called cytokines, can interact with certain immune cells, to produce a cell that is "primed", that is the cell is sensitised and potentially hyperactive as compared to corresponding non-primed cells. It is a little like a balloon which has lots of air in it and, when pricked by a tiny pin, will release the air in a sudden rush, as compared to another balloon containing little air. Continuing the analogy, a primed cell, like the balloon, is ready to go and all it takes is the pin or, in the case of a primed immune cell, some other substance to produce an effect, a bit like the balloon. There are examples of pollutants, such as pesticides that, at non toxic levels, can greatly augment the response of immune cells. Studies carried out by a group of Italian scientists showed that chronic exposure rat immune cells to permethrin, a well known pesticide, at levels not considered to be toxic, primes the cells and this results in a greatly amplified response to other stimuli [4]. The cells studied by the researchers produce free radicals (mentioned earlier in this book) which can kill bacteria but, in excess, free radicals can also damage delicate tissue causing disease such as arthritis. The permethrin-primed response measured by the scientists was so great (more than 30 times greater than normal) that the researchers speculated that, if something similar occurred in humans, then chronic exposure to some pesticides had the potential to harm. A similar augmentation of an immune response has also been shown in laboratory mice [5]. In this case the mice had been previously primed with albumin, a protein found in blood, which had rendered their lungs very sensitive (a type of priming). Motorcycle exhaust particles, known to contain a variety of pollutants with the capacity to induce a type of lung inflammation, were introduced into the lungs of both sensitised and non-sensitised animals and the response to these treatments was measured. The researchers concluded that prior sensitisation and subsequent treatment with a mixture of environmental pollutants greatly augmented inflammatory processes in the lung. This is especially relevant to people with asthma, whose airways are particularly sensitive.

What Does All of This Mean?

It has been generally assumed by governments that, because we are only exposed to tiny amounts of environmental chemicals and pollutants it is unlikely that this exposure can cause us harm. This view may be flawed because there is ample evidence that, in the case of essential nutrients, for example certain vitamins, the amounts required to prevent disease are in the order of millionths of a gram. Also, we know that sub-microgram amounts of some substances, most notably certain bacterial poisons, can kill us, and very small amounts of certain food proteins can cause severe allergic reactions and even death so small can be lethal. We may need to re-evaluate our beliefs on toxicity because they are often based on animal studies that do not take into account differences between animals and humans, and the fact that much of our exposure can occur over many years [6]. There is increasing evidence that exposure to tiny amounts of a toxic chemical may have unexpected effects through the process of hormesis, or through synergism or priming which depend on the combined actions of a pollutant with other substances that may be present in our blood and tissues. Some of the pollutants we are exposed to have been demonstrated to produce effects via these processes in animal, and even human, tissues such as the brain and the immune system. However, there are literally thousands of chemical reactions occurring in our bodies at any time and it is likely that at least some of these reactions are either inhibited or stimulated in the presence of small amounts of one or more of the many pollutants taken up into our bodies, possibly affecting the function of the organ(s) in which the reactions are taking place.

References

1. Jiang G et al (2009) Biphasic effect of cadmium on cell proliferation in human embryo lung fibroblast cells and its molecular mechanism. Toxicol In Vitro 23:973–979
2. Velini ED et al (2008) Glyphosate applied at low doses can stimulate plant growth. Pest Manag Sci 64:489–496
3. Gao HM et al (2003) Synegtistic dopaminergic neurotoxicity of the pesticide rotenone and inflammogen lipopolysaccharide: relevance to the etiology of Parkinson.'s disease. J Neurosci 23:1228–1236
4. Gabienelli et al (2009) Effect of permethrin insecticide on rat polymorphonuclear neutrophils. Chem Biol Interact 182:245–252
5. Lee CC et al (2008) Motorcycle exhaust particles augment antigen-induced airway inflammation in BALB/c mice. J Toxicol Environ Health A 71:405–412
6. Gwinn MR, Axelrad DA, Bahadori T, Bussard D, Cascio WE, Deener K, Dix D, Thomas RS, Kavlock RJ, Burke TA (2017) Chemical risk assessment: traditional vs public health perspectives. Am J Public Health 107(7):1032–1039. https://doi.org/10.2105/AJPH.2017.303771. Epub 2017 May 18

Chapter 20
What Can We Do?

There is really very little doubt that, with continuing globalization, and with the emergence of the two most populous countries in the world, India and China, as industrial powers with the potential to rival the US and Europe, chemical exposure of the planet's seven billion people will increase. The rush to industrialization is really unstoppable as almost a half of the world's population now aspires to a better standard of living. Unfortunately, this rush to industrialization has consequences for all of us. While the focus has been largely on the release of carbon dioxide and its impact on the Earths's climate, it seems that the corresponding release of many other pollutants has not attracted the same attention. Perhaps the main reason for this is the belief that there is no evidence that the many chemicals to which humankind is now exposed are harmful. After all, so the argument goes, our governments have put in place processes to regulate our exposure to chemicals. Moreover, even if we are exposed, our governments have essentially been reluctant to act firstly because of the impact on growth and jobs and, secondly, because they remain unconvinced that the amounts to which we are exposed are sufficient to cause harm. The Government's response has been to reverse the onus of proof ie we do not think that there is a problem but, if you do, prove it. Of course, if something does produce an acute effect, then the link between exposure and disease is easy to prove. However, if any effect is not immediate but requires a constant exposure over many years, it is exceedingly difficult to prove. And, without this proof, governments will not normally act.

Now we know that governments have acted in the past to ban the use of certain chemicals even though there was not incontrovertible evidence of exposure and disease. The PCBs are a good example of this. What happened in this particular case is a good illustration of the time taken for governments to act. The PCBs were first developed in the around 1930 as an insulating material with excellent stability to alkali, acid and heat. Its uses multiplied and thousands of tons were manufactured to cope with these diverse needs. As the years went by there were occasional reports of possible effects on animals, then there were accounts of the detection of PCBs in marine animals, the surprising environmental stability of PCBs, their detection in the food chain and in human tissues, and possible effects on humans. Despite all of the accumulating evidence of a build up in animals that were part of our food chain, and the reported harmful effects on animals, it was not until 1979, fifty years after they were first used, that their manufacture was banned in the USA. It is most instructive to read the US Environmental Agency report on the reasons for the ban which included the following statements - "PCBs have caused birth defects and cancer in laboratory animals, and they are a suspected cause of cancer and adverse skin and liver effects in humans. EPA estimates that 150 million pounds of PCBs are dispersed throughout the environment, including air and water supplies; an additional 290 million pounds are located in landfills in this country". Of course that was not the end of the story because, as even the EPA acknowledged that there still remain "a vast amount of PCBs still in use" which had to be disposed of in some way, via landfills or incineration. Much like trying to change the course of the Titanic, it takes time and, even if the manufacture is banned, the environmental impact can be felt for many years after a decision is made. Thus it is that, thirty years after the banning of the manufacture of PCBs in the US, it is still being detected in human tissues and body fluids including in the inhabitants of some of the most remote parts of the planet.

It is too easy to invoke a 'conspiracy theory" to explain why it took governments so long to act. While multinational – government collusion is a favoured theory of some, the more likely explanation is that governments were not convinced of the potential harmful effects of PCBs.

So, given that that we are now exposed to thousands of environmental chemicals, only some of which have been mentioned in this book, and given also that we now know that some of the environmental chemicals have already been shown to be harmful, and there are literally thousands of others we know very little about and we also know nothing about how they interact with each other, what can we do? Are we completely powerless in the face of what appears to be a "chemical armageddon"?. While there is a lot we can do on a personal level, and more about that later, it really is a national problem and therefore falls within the jurisdiction of individual governments. Indeed, with the seas and marine life gradually filling with plastics and industrial waste products, and our food chain becoming increasingly contaminated wherever we live, it is an international problem and one that should be debated by the UN/WHO. However, the difficulties we face in getting governments to recognise the severity of the problem is clear from the recent Paris meeting on climate change. It seems that governments do not take a threat seriously unless it is immediate, as is evident from the melamine milk scare in China and the clenbuterol episode in Italy, and they will then act. Part of the reason for this no doubt is the fact that there have been lots of doomsday scenarios predicted in the past which have not eventuated so the argument goes – why should this be any different? Also, any action taken by a government can have a major impact on some groups, which can mean loss of jobs and profits for the community, and loss of electoral support for the government. If individual governments find it difficult to respond quickly, it is even more difficult reaching an international consensus particularly as the developing world is striving to improve the living standards of its huge populations and to do this there is no doubt there will be a reliance on those resources that will add to the global chemical burden.

It is easy to be pessimistic about prospects of cutting global emissions when even the developed countries seem unable to reach a consensus on how to do this but pessimism almost certainly guarantees failure. There really is no easy solution but to put electoral pressure on individual governments. It is worth reminding ourselves that all of the political and social changes that have altered peoples' lives have occurred via this process so there are many precedents. But it will take time and levels of pollutants will increase inexorably. At present there is a belief that there is no urgency and we can take all the time we want. What we do not know at present is what the consequences are of the big increase in pollution of the seas, the air, and the food chain that will occur in the next decade or so. It has always been assumed that whatever waste we produce will eventually be degraded either by the myriads of microorganisms that are present in the environment or by the normal non-biological processes that are able to wear down even objects like rocks and stones that are seemingly indestructible (ie heat, wind, ice, water and UV) rocks. It is true that nothing is forever but some things take a long time to degrade and, because they do, they will accumulate, particularly in the soil and sea. Good examples include non-biodegradable plastics, organochlorines such as DDT and other pesticides and industrial waste products such as lead, mercury, PCBs, and PBDEs. This has already happened as is apparent from the presence of some of these pollutants even in marine life that is a long way from where they were generated and released into the environment. What has also become apparent in the last few years is that even if the amounts of pollutants may not be large at the moment, they can build up to quite high levels via biomagnification, a process that has been mentioned earlier in this book, so much so that governments have become alarmed and have cautioned particularly vulnerable sections of the community, in particular pregnant women, to reduce their intake of larger predator fish because of their increased levels of pollutants.

There are steps that individuals can take to reduce their exposure to environmental pollutants and thereby reduce the risk of degenerative diseases such as cancer. As food is a significant source, and many of the pollutants are introduced at various stages of food production, from growing the crops, to storage, transport and packaging, it makes sense to buy or grow food that is low in foreign chemicals. This can be done by growing one's own food, or buying organic or biodynamic food. At the same time, by doing this, we are encouraging the use of sustainable, non-chemical processes for growing, storing and transporting food. As processed food is often stripped of essential nutrients, chemicals such as preservatives are added, and processes are used which can generate other, potentially harmful substances, we would be best served if we chose fresh food wherever possible although simple, well tried methods of preservation such as drying or freezing are unlikely to contain additional chemical contaminants. As plastic food packaging is yet another source of contaminants, it is preferable to use paper or glass to store food, particularly food that has a significant fat content. Drinking water is yet another source of contaminants. Ground water from which drinking water is derived often contains small amounts of a variety of environmental contaminants, sometimes referred to as "organic matter", which have found their way into the water via different routes. This organic matter is further subjected to chlorination which generates yet another group of substances which have been referred to as "chlorination byproducts". As there are indications that some of these substances are potentially harmful, filtration of water to remove much of this organic material and the chlorination byproducts could be considered. Alternatively, collection of rainwater in tanks is another option although environmental contaminants can find their way onto the collection surface from air pollution and then into the water. Chemicals in the collection surface (most the roof of a house) and the storage tank can also contaminate tank water, as can materials from bird droppings and decaying leaves so filtration is worth considering.

It is difficult to avoid air pollution for people who live in cities, particularly in high density areas. Clearly, the move to renewable energy, and the corresponding phasing out of energy production from fossil fuels, should be encouraged, as should the development of electric motor vehicles. As this will take time, moving to lower density housing areas, perhaps even semi rural areas on the outskirts of a city, could be an option although there are clearly disadvantages in this as there are increased costs and time associated with commuting the extra distance from home to work. The use of chemicals in the home and garden eg insecticide sprays, air deodorisers (sometimes attached to power outlets), heavy duty detergents and shampoos, could be restricted. To reduce the exposure to indoor chemicals, the use indoor purifiers could be considered. As our place of employment is often a source of chemical exposure, it would be worthwhile finding out from management what chemicals are used and the precautions taken to limit this exposure if there are any indications that any of these are potentially harmful. Websites showing lists of carcinogenic chemicals have been mentioned in an earlier chapter are worth consulting.

We are living in a time when the world population is far greater than it has ever been, and perhaps even envisaged, and where a half or more of that population is poised to greatly increase its standard of living. To do this will require the use of more of the earth's resources than ever before. At the same time, chemical pollutants will be released into the environment on a far greater scale than at any other time in human history. There is already clear evidence that literally no place on earth, even the most pristine environments that are far removed from industrial activity, are free of pollutants. It is also clear that they find their way into the water we drink, the air we breathe and into the food chain and from there into our bodies in not insignificant amounts. We also know that environmental factors, including chemical pollutants, already contribute to the development of a variety of diseases, including respiratory diseases, cancer, diabetes, heart disease, multiple sclerosis, and arthritis to list just a few. As both the diversity and quantity of pollutants increases in the future in step with a burgeoning population, it will be indeed surprising if there is no corresponding impact on our health and wellbeing. It is hoped that governments recognise that pollution is a problem with potentially catastrophic consequences for humankind and that action is taken before it is too late to reverse.

Index

A

Acceptable dietary intake (ADI), 11
  assessment of potential toxicity, 12
  maximum residue limits and, 12
Acetylcholine
  action of neonicotinoids on release of, 7
  effect of acetylcholinesterase on, 25
  effect of organophosphates on, 7
  release from nerve cell, 25
Acetylcholinesterase, 25
  effect of organophosphates on, 7
  presence in insect, rodent, and human
    brain, 25
Aerosols, 76
  air pollution and, 76
  presence in diesel exhausts, 136
Agricultural and industrial chemicals in, 100,
    104
Air pollution
  aerosols as components of, 76
  cancer risk and effect of, 80
  cardiovascular disease and effect of, 80
  composition of, 75–76
  effect on lung function, 80
  fossil fuels and production of, 75
  gases as components of, 75–76
  particulate matter, presence in, 76
Aldrin, 7, 16
Aluminium phosphide, 19
Alzheimer's disease, 3
Antibiotics
  antimycin and mitochondria, 212
  bacterial diseases in plants and use of, 7
  cyclosporine and organ transplants, 217

oligomycins and mitochondria, 212
Aromatic amines
  dye and chemical industries and, 137
  occupational exposure to and bladder
    cancer, 137
Arsenic, 55–56
  cancer and, 134
  dimethylarsenic, 57
  effects on DNA, 207
  effects on health, 56–58, 133
  occupational exposure to, 133
  presence as contaminant in milk
    powder, 57
  presence in coal, 56
  presence in food, 56
  presence in ground water, 56
  susceptibility of young children, 57
  uses of, 6, 55, 133
Arsenic in, 86, 91, 97–99, 108
Asbestos
  mesothelioma and exposure to, 130, 137
  occupational exposure to, 127
Asthma
  diacetyl exposure and, 136
  exposure to air pollution and, 78
  exposure to formaldehyde, 134
  exposure to pesticides and, 133
  sulphur dioxide and, 78
Australian Pesticide and Veterinary Medicine
    Authority (APMVA), 12, 13
Autoimmune disease
  lupus, 3
  multiple sclerosis, 3
  rheumatoid arthritis, 3

B

Baciillus thuringiensis (Bt), 7
Batteries, 182
Benzene
    exposure and leukemia, 135
    industrial exposure to, 135
Benzophenones, 44
Benzopyrene, 207
Beryllium, 134
    occupational exposure and fibrosis, 134
Bioherbicides, 10
Biopesticides, 7
Birth defects
    birth defects in four children, 20
    chlorpyrifos exposure and, 20
    thalidomide exposure and, 191
    valproic acid exposure and risk of, 191
Bisphenol A
    heart disease and exposure to, 43
    insulin and effects of, 43
    puberty, 42
    sexual development and effects of, 42
    sperm counts and effects of, 42
    thermal papers component, 122
Bisphenol F, 43
Bisphenol S, 44
Blood brain barrier, 26
Bottled, 89–90, 104
Breast cancer
    BRCA-1 and BRCA-2 and risk of, 201
Breast milk
    fluorocarbons presence in, 150
    mercury presence in, 54
    PBDE presence in, 70
    PCBs presence in, 64, 65
    pesticides presence in, 24

C

Cadmium
    exposure in animals and cancer, 132
    hormesis and effects of, 225
    kidney function and exposure to, 133, 225
    rock phosphate fertilizer and contamination
        with, 132
    silver jewellery production exposure to, 132
Cadmium in, 86, 91, 99
Cancer
    air pollution and, 81
    aromatic amines and, 137
    arsenic and, 57
    benzene and, 135
    BRCA-1 and BRCA-2 and, 190

    cadmium and, 132–133
    chemicals and, 191
    chromium and, 134
    diesel exhaust and, 136
    dioxins and, 67
    PCBs and, 66, 67
    pesticides and, 21
    polymorphisms and risk of, 201–202
    radioactivity/radiation and, 172–175
    TCDD and, 121
    uranium mining and, 130
    vinyl chloride and, 137
    Woodworking and, 131
Carbon dioxide
    air pollution and, 76
    climate change and, 1
    global warming and, 1
Carbon monoxide
    death by poisoning from, 79
    formation of carboxyhaemoglobin
        from, 2, 79
    paper manufacture and formation of, 117
Cardboard
    chemical migration into packaged food
        from, 123
    chemicals in recycled paperboard, 123
Chemical contaminants in, 97, 99, 100
Chlordane, 7
Chlorination and
    trihalomethanes, 106
Chromosomes
    autosomal dominant genes, 199
    autosomal recessive diseases, 198
    autosomes, 198, 199
    cystic fibrosis, 198
    Huntington's disease, 198
    mutations on, 199
    phenylketonuria, 198
    X and Y chromosomes, 198
    X linked genetic diseases, 199
Cigarettes
    presence of chemicals in smoke, 223
    smoking and lung cancer, 224
Coal
    combustion and release of metals, 49–50
    Federal Coal Mine and Safety
        Act and, 129
    heavy metal contamination, 49
    mining and black lung disease, 129
Cockayne syndrome, 205
Codex Alimentarius Commission, 2
Combined immunodeficiency, 216
Cypermethrin, 8

D

Degradable plastic, 36–37, 39
Deltamethrin, 8
Derris dust, 8
  rotenone and Parkinson's disease, 20
  rotenone presence in, 212
Desalination of, 85, 90–91
Diacetyl, 135–136
  inflammation of nasal passages and
    exposure to, 136
  popcorn makers and exposure to, 136
Dibutylphthalate
  transgenerational effect of, 42
Dieldrin, 7
Diesel exhaust
  chemical composition of, 136
  composition of, 80, 136
  formation of thrombus from, 80
  mining exposure and lung cancer risk from,
    136
Diisothiocyanates, 137
Dimethoate, 7
2, 4-Dinitrophenol, 212
Dioxins
  bleaching of pulp and formation of, 117,
    119
  cancer risk from exposure to, 67
  incineration and production of, 119
  seveso and release of, 121
  toxicity of, 117
Disinfection byproducts
  and bladder cancer, 107, 108
  and toxicity, 103–104, 108
Disinfection of, 86, 89, 90, 92, 95–99
DNA
  benzopyrene effects on, 207
  bisphenol A and methylation of, 208
  damage to, 206–208
  demethylase, 208
  dioxin exposure and, 207
  diethylstilbestrol and transgenerational
    effects, 209
  epigenetic changes to, 209
  8-HDG and breakdown of, 207–208
  histones, 208
  methylation of, 208
  micronuclei formation, 207
  phthalates effects on, 209
  PBDE 47 effects on, 207
  PCB effects on, 207
  reaction with polycyclic aryl hydrocarbons,
    207
  repair of, 208

Drinking water
  and disease, 99, 100

E

Endosulfan, 7, 16
Environmental Protection Agency (EPA)
  birth defects and, 20
  carcinogenic pesticides and, 21
  regulation of pesticides, 12–13
Epigenetic
  bisphenol and phthalates effects, 209
  diethylstilbestrol and transgenerational
    effects, 209
  from exposure of fathers, 209
  methylation and demethylase, 208
  micro RNA and histones, 208
Environmental release of chemicals
  from, 182–183

F

Farmers/agricultural workers/pesticide users
  cancer risk in, 21–22
  lung and respiratory function, 131
  occupational exposure to pesticides, 129–
    130
  organic food intake of, 22–23
  Parkinson's disease and pesticide exposure,
    20
  pesticide users and risk of prostate cancer,
    22–23
  semen quality and, 22–23
Fertilizer, 132
Fire/flame retardants
  particulate matter containing, 78
  plastics containing, 36, 37
  PBDEs in, 68
Fenvalerate, 8
Fish oils, 190, 217
Fluorocarbons
  breast milk presence of, 150
  daily intake in food and water of, 145
  drinking water contamination by, 145
  gestational diabetes and exposure
    to, 151
  heart and kidney disease and exposure to,
    150
  indoor air and household dusts presence
    of, 146
  LD50 of PFOA and PFOS, 147
  migration into food by, 145
  presence in human tissues of, 145

Fluorocarbons (*cont.*)
    rats and monkeys effects of, 147
    semen quality and birth weight effects of,
        149
    thyroid function effects, 151
Food handling
    baker's asthma, 132
    flour dust, 132
Food Standards Australia New Zealand food
        safety monitoring by, 13, 14
    Total Diet Study, 13, 14
Formaldehyde
    enhancement of response to allergens by,
        218
    exposure and cancer, 135
    immune system effects of, 218
    indoor air pollution and, 77
    occupational exposure to, 135
Fossil fuel, 75, 77
Fukushima, 172, 174
Fumigants
    eye and skin irritation and exposure to, 19
    respiratory problems and exposure to, 19
    shipping containers and exposure to, 9
Fungicides
    crop infections treated with, 9
    mode of action of, 9
    natural killer cells and, 218

G
Gas pollutants, 78–79
Genetic polymorphism, 199–203
    breast cancers, 201
    cytochrome P450, 200
    excretion of environmental chemicals and,
        201
    heart disease risk, 202–203
    Maxolon and dystonia, 200
    statins and drugs, 200
Genetic variability, 197–204
    drugs and, 200–201
    twin studies, 201
Global warming, 1
    Kyoto Agreement, 1
Glyphosate
    antibiotic effects and gut bacteria, 22
    formulations, 28
    hormesis, 225
    toxicity, 18
    well water exposure and kidney
        disease, 22

Glutaraldehyde, 137
Government regulation of, 97–98
Graywater, 88, 92
Groundwater, 87, 91, 99, 101, 108

H
Health effects of, 183–185
Hearing
    cochlea and, 157
    ear and, 157–159
    ototoxic chemicals and, 161, 162
Hearing loss
    agricultural chemicals and, 161
    cisplatin, 161
    heavy metals 161
Heart disease
    homocysteinuria and, 203
    hypercholesterolemia and, 203
    non genetic factors and, 202, 203
    saturated and trans fats, 202
    Seventh Day Adventists and risk of, 203
    Vegetarians and risk of, 203
Herbicides
    bioherbicides, 8
    classes of, 8
    glyphosate, 18
    2, 4, 5 T and TCDD release, 121
Hormesis, 225
    cadmium and low dose stimulation, 225
    glyphosate and low dose stimulation, 225
Household air pollution, 76
    biomass burning and, 76
    cooking and, 76
    volatile organic components (VCOs) in, 77
Hydrocarbons
    adaptive immune system, 216, 218
    adjuvants in vaccines, 217
    AIDS and HIV, 217
    anaphylaxis and peanuts, 217
    apoptosis, 218
    aromatic, 135
    B-cells and T-cells, 215, 218
    combined immunodeficiency, 216
    chronic granulomatous disease, 216
    chronic inflammation, 216, 220
    cytokine production by, 215
    diesel exhaust and T-cells, 218
    fish oils and inflammation, 217
    flagellin and stimulation of, 217
    hydroquinone and macrophages, 219
    immunosuppression and steroids, 224

innate immune system, 215–216, 219
motor vehicles and, 79
macrophages, 215, 219
macrophages and pesticides, 219
marine mammals and effects of pollutants, 219, 220
natural killer cells and pesticides, 218
neutrophils, 215, 216
PCBs and T-cells, 218
petrol and diesel components, 136
pollutants and toxicity of immune system, 219–220
priming, 226–227
superoxide and neutrophils, 216
tuftsin, 217

I

Indium, 134
Indoor air pollution biomass and, 77
formaldehyde, 77
volatile organic components (VOCs), 77
Insecticides
organochlorines, 7–8, 21, 26
organophosphates/organophosphorus, 7, 17, 27
neonicotinoids, 7

K

Kidney
disease and arsenic, 57
disease and pollutants, 150–151
drugs and damage to, 200
environmental chemicals and urine elimination, 200
transporters and excretion by, 200
Kyoto agreement, 1
Kyshtym nuclear processing site, 168

L

LC50, 10
LD50, 10–12, 121, 146, 224
Lead
atypical Parkinson's disease and exposure to, 53
ayervedic medicines and presence in, 52
blood safe levels of, 52
canning and release of, 51
health effects of, 52, 53
lead tetraethyl and petrol, 50
occupational exposure to, 51
Port Pirie and smelting, 51
presence in coal, 49
presence in food, 51
Lead in, 182–184
Leukemia
air pollution and, 80
Chernobyl liquidators, 173
formaldehyde exposure and, 134–135
genetic component and risk of, 202
low dose radioactivity exposure and risk of, 171
pesticide exposure and risk of, 21
Life span, 190
Sardinia and Okinawa and centenarians, 190
Lipopolysaccharide (LPS), 226
rotenone and synergism, 226
Liver
cytochrome P450, 192, 200
elimination of environmental chemicals, 192–193
impaired function and susceptibility to pesticides, 27
Lupus, 3, 191
Lymphoma
benzene exposure and, 135
dioxin exposure and, 67
pesticide exposure and, 23

M

Malathion, 7
Marine mammals, 219
Maximum Residue Limits (MRL), 12, 16
Mercury, 53–55
biomagnification in fish, 54
birth defects and pregnant women, 55
breast milk and presence of, 54
fungicide and treatment of grain, 54
Minamata Bay and release, 54, 55
occupational exposure, 55
presence in coal, 49, 53
Mesothelioma, 130
Metals in, 182–185
Metamsodium, 19
Methomyl
contamination of food, 18
Methyl bromide, 19
Mevinphos, 7
Micronuclei, 207
in electronics workers, 207
from exposure to air pollutants, 207

Micronuclei (*cont.*)
    from exposure to bitumen fumes, 207
    from formaldehyde and pesticide exposure,
        207
Microplastics, 38
Microwave syndrome, 175
Mining black lung disease and, 129
    asbestosis and, 129, 130
    silicosis and, 130
Mitochondria
    ATP production by, 211
    derris dust effects on, 212
    DNA in, 212, 213
    effects of environmental chemicals on,
        211–213
    electron transport chain of, 211
    fluorocarbons and abnormalities in function
        of, 219
    free radical formation by, 211
    inherited abnormalities in, 212
    rotenone and paraquat effects, 212
Mobile phones, 176
Motor neurone disease, 189
    PCB exposure, 66
    pesticide exposure, 20
MPTP, 191
Multiple chemical sensitivity
    pesticides and, 27
    toxicant-induced loss of tolerance (TILT),
        27
Multiple sclerosis, 3, 191

N
Nanomaterials, 38
National Residue Survey, 13
Neem, 8
Neonicotinoids, 7
Neural tube defects
    pesticide exposure and, 26–27
Nitrogen dioxide
    effects on asthmatics, 79
    formation from fossil fuels, 75, 79
    heart function and, 79
Noise
    acoustic neuroma and, 60
    cardiovascular disease and, 159–160
    genetic factors and, 161
    hearing loss and, 159
    occupational, 159, 161
    sleep and, 160
    solvents and, 161

    stress hormones and, 159
    tinnitus and, 157, 160–161
    weight gain and, 160
Non degradable plastic, 34
Non-metallic occupational exposure, 134–137
    benzene, 135
    diacetyl, 135–136
    formaldehyde, 134–135
No observed adverse effect levels (NOAEL), 11

O
Occupational exposure and
    asthma, 129, 132, 135–137
    chronic obstructive lung disease (COPD),
        129
    dermatitis, 130
    lung diseases, 129, 134, 137
Occupational metal exposure effects, 132–134
    arsenic, 133
    beryllium, 134
    cadmium, 132–133
    chromium, 134
    indium, 134
Organic food /diet
    Danish farmers and organic food intake,
        22–23
    pesticide levels in urine, 30
    pregnant women and pre-eclampsia risk, 23
Organochlorines, 7, 8
    presence in brain, 193
    presence in polar bear tissues, 219
    stability of, 8, 64, 194
Organophosphates/organophosphorus, 7
Ototoxic chemicals
    hearing loss and, 161, 162
Ozone
    health effects, 79
    presence in air pollution, 75
    treatment of water with, 79

P
Paper manufacture, 115–119
    additives and chemicals added in, 118
    chemical byproducts of, 116–118
    chlorine and bleaching, 116, 117, 119
    generation of trihalomethanes, 120
    hypochlorous acid and, 119
    non chemical processes, 118
    pulp mill effluents and fish, 120
    pulp mill waste and chemicals, 118–119

Paper waste
    landfills and release of chemicals into soil
        from, 122
    recycling and generation of chemicals from,
        122
Paraquat, 8, 212
Parkinson's disease
    agricultural and horticultural workers and
        risk of, 20
    MPTP exposure and, 212
    PCBs exposure and, 66
    pollutants in substantia nigra and, 193
Parkinson's disease and exposure to, 66
    presence in breast milk, 65–67
    presence in food, 65, 66
    rice oil contamination wit, 66
    stability, 64
    thyroxine effects, 66
    uses of, 64
    Yucheng and Yusho contamination, 66
Parkinson's disease and exposure to, 26
    residues in food, 16–19
    Total Diet Study, 16, 18
    Toxicant induced loss of tolerance (TILT),
        36
    varying susceptibility to, 31–36
Particulate matter, 76–80
    chemical composition of, 76–77
    lung inflammation and, 79–80
    presence in food and water, 78
    presence in lungs, 77–78
    solids and bound material in, 76–77
    urine excretion of chemicals from, 78
PBDE 47 and DNA damage, 207
    TSH reduction and, 70
Perfluorooctane sulphonate (PFOS), 144
    birth weight of babies and, 150
    chronic exposure to, 148
    effects on rats and monkeys, 147
    environmental release of, 144
    heart disease and, 150
    LD50 of, 147
    presence in food and water, 145
    presence in human tissues, 145
    thyroid disease and, 151
    uses of, 144
Perfluorooctanoic acid (PFOA), 144
    birth weight of babies and, 150
    chronic exposure to, 148
    effects on rats and monkeys, 147
    environmental release of, 144
    heart disease and, 150
    LD50 of, 146

legal class action and, 144
    presence in food and water, 145
    presence in human tissues, 145
    thyroid disease and, 151
    uses of, 144
Permethrin, 8, 227
Pesticides
    amyotrophic lateral sclerosis and exposure
        to, 20
    asthma and respiratory problems and
        exposure to, 19
    cancer and childhood cancer, 21, 22
    chronic exposure to, 18–19
    detoxification by paraoxanases, 192
    exposure and birth defects, 20
    foetal uptake of, 20
    formulation and polyoxyethyleneamine, 28
    government regulation of, 12–17
    liver function and, 27
    occupational exposure to, 19, 132
    organic food and levels of pesticides, 22–23
Pharmaceuticals/drugs
    blood levels and cytochrome P450 effects,
        200–201
    cytochrome P450 and kidney damage, 200
    cytochrome P450 and polymorphisms,
        200–201
    Thalidomide and teratogenic effects, 191
Pheromones, 8
Photovoltaic cells, 182, 185
Phthalates
    breast development and, 41
    cryptorchidism and, 41
    endocrine disrupting effects of, 41, 42
    heart disease and, 43
    hypospadias and, 41
    insulin resistance effects of, 43
    obesity and, 43
Plastic degradation
    biodegradable, 36, 37
    burning, 39
    landfills and, 39
    non degradable, 39
Plastics, 33–44
    additives, 37–38
    antioxidants, 37, 40, 44
    burning of plastics, 39
    degradable, 36, 37
    migration of additives into food and water,
        40
    nanomaterials in, 38
    presence of additives in breast milk, 40
    stability of, 33, 34, 36

Plastics (*cont.*)
  uptake into the body of components of,
    39–41
Plastics Europe, 33
Plastics in, 101–102
Plumbing contaminants
  lead and, 102
Polar bears and pollutants, 219
Polyamides, 35
Polybrominated biphenyls (PBBs), 70
  Michigan cattle feed contamination and, 70
Polybrominated diphenyl ethers (PBDEs),
    68–71
  breast milk and presence of, 70
  Chinese workers and TSH effects, 70
  effects on physical and mental development,
    70
  effects on thyroid hormones by, 69, 70
  food chain and presence of, 68–70
  presence in body fat, 69
  presence in breast milk of, 69, 70
  presence in food chain of, 68, 70
  stability of, 68
  uses of, 68
Polycarbonate, 35
Polycarbonate, 35
Polychlorinated biphenyls (PCBs), 64–68
  changes in DNA by, 207
  chloroacne and exposure to, 66
  dioxin formation from, 65
  furan production from, 64
  infant development effects of, 67
  male reproduction effects of, 67
  menstruation effects, 66
  motor neurone disease and exposure to, 66
Polyethylene, 35, 36, 39
  chemicals in, 40
  migration of chemicals into food and water
    from, 40
  polyethylene terephthalate (PET), 35
Polylactate, 36
Polymerisation
  initiators and, 34
  monomers and, 34
Polyoxyethyleneamine (POEA), 28
Polypropylene, 35
Polystyrene, 35
Polyvinyl chloride, 35
Priming, 226–227
  cytokines and immune cells, 216
  motor cycle exhaust particles and, 227
  permethrin and albumin and, 227
Pyrethrins, 8

R

Radiation/radioactivity, 166–177
  accidents and release of, 173–175
  atomic bomb survivors, 173, 177
  cataracts and exposure to, 175
  Chernobyl and liquidators exposure to,
    173–175
  dosage and health, 170–173, 175–177
  Fukushima and release of, 174
  iodine-131 and release of, 172, 174–175
  microwave syndrome, 175–176
  mobile phones, 176
  MRI, 176
  Nagasaki and Hiroshima, 171–173
  radioactive elements used and, 167
  sources of, 166, 168–169
  Three Mile Island and release of, 168
  thyroid cancer and, 172, 173, 175
  WiFi and, 176
Radioactivity in, 101
Rainwater, 89, 90
Rare earths in, 182–185
Recycled, 89, 92
Recycling
  paper and paperboard, 122
Recycling of, 182–185
Regulation
  Australian Pesticide and Veterinary
    Medicine Authority (APVMA) and
    pesticides, 12, 13, 15, 16
  Environmental Protection Agency (EPA)
    and pesticides, 12, 13, 15
  Food Standards Australia and New Zealand
    and food (FSANZ), 13–15
  Therapeutic Goods Administration (TGA)
    and drugs, 15
Reservoirs, 86–87, 92–95, 97, 100, 101
Rheumatoid arthritis, 3
  exposure to pollutants and risk in women,
    220

S

Semen/spermatozoa
  fluorocarbons and, 149
  PBDEs and, 69
  PCBs and, 67
  phthalates and, 41
Seveso
  explosion and release of TCDD, 121
  exposure to TCDD and chloracne, 121
Silica
  inflammatory response to, 129

occupational exposure to, 129
   silicosis and mining, 130
Solar panels, 182
Sound
   intensity, 158, 159
   sources, 158
Sources of, 184
Squalene, 217
Stormwater, 87, 92
Sulphur dioxide
   acid rain and formation from, 78
   effects on asthmatics of, 78
   presence in air pollution, 76, 78
   preservative use, 78
Susceptibility, 24–27
   children and blood brain barrier, 24–26
   genetic polymorphisms and, 199–203
   liver function and, 27
   unborn baby, 26
Synergism, 226
   pesticide mixtures and immune cells and,
      226
   rotenone in derris dust and LPS, 226

T
Tank water, 89
Tapwater, 91, 102
Teflon, 35
Teratogenic effects
   thalidomide, 191
   valproic acid, 191
Total Diet Study, 13, 14, 51
Thyroid/Thyroxine
   effects of dioxins on, 67
   effects of PCBs, 67, 68
   effects of PBDEs, 70
   fluorocarbons and, 151

iodine-131 and, 172, 174
Toxicant induced loss of tolerance (TILT), 27
Transport and distribution, 97
Tumour suppressors, 206
   Li-Freumeni syndrome, 206
   P53 gene, 206
Twins
   dizygotic and monozygotic, 201
   lung cancer risk of, 202

U
Uranium mining, 130
   lung cancer from, 130
   radon exposure from, 130
US Food and Drug Administration (FDA),
   13–15, 43

V
Vinyl chloride, 35
   animal carcinogen effects of, 138
   liver cancer and exposure to, 137
   monomer of polyvinylchoride (PVC), 137

W
WiFi, 176
   blood pressure and exposure to, 176
   heart rate and exposure to, 176
Windmills, 182
Woodworking, 131
   IARC and carcinogen, 132
   wood dust and cancer, 132

X
Xeroderma pigmentosum, 205

Printed in the United States
by Baker & Taylor Publisher Services